George King

The theory of finance

being a short treatise on the doctrine of interest and annuities certain

George King

The theory of finance
being a short treatise on the doctrine of interest and annuities certain

ISBN/EAN: 9783743335547

Manufactured in Europe, USA, Canada, Australia, Japa

Cover: Foto ©berggeist007 / pixelio.de

Manufactured and distributed by brebook publishing software
(www.brebook.com)

George King

The theory of finance

The Theory of Finance:

BEING

A SHORT TREATISE ON THE DOCTRINE OF INTEREST AND ANNUITIES-CERTAIN.

BY

GEORGE KING, F.I.A.

OF THE ALLIANCE ASSURANCE COMPANY.

PREFACE.

THE following treatise consists of Notes of Lectures on Interest and Annuities-Certain, delivered to the Students for the Intermediate Examination of the Institute of Actuaries. The course of lectures includes also the mathematical theory of Life Contingencies, and the author's intention originally was, to have edited and published the whole of his notes under the title of the *Elements of Actuarial Science.* The absence of such a guide is very much felt by students; and although it has long been the declared intention of the Institute to bring out a complete Text-Book, yet it was thought that a less pretentious work, speedily published, would in the meantime fill up the gap, while afterwards it might serve as a useful introduction to the more elaborate volume.

At the close of last year the author accepted the invitation, with which he was honoured by the Council of the Institute of Actuaries, to prepare that part of their Text-Book which is to treat of the Science of Life Contingencies, and his original intention therefore fell to the ground. He thought, nevertheless, that, as the five chapters on Interest and Annuities-Certain were almost finished, and could be made, with but slight alterations, complete in themselves, they might not inappropriately be published independently, and he offered the manuscript to the Actuarial Society of Edinburgh. He feels much gratified and flattered at the cordial way in which his proposal was accepted, and he desires to express his hearty thanks to the Committee of that Society for the uniform and courteous consideration with which they have received his suggestions. He hopes that the end both they and he have in view may be secured, and that students more particularly, but also the actuarial profession generally, may derive benefit from the labour which was undertaken without the prospect of any pecuniary return.

The available space being very limited, much has been omitted which, under more favourable circumstances, would have been, and with great advantage, included. An effort has been made to bring this branch of actuarial science up to the latest date, but those formulas and methods which are now more of historical than practical interest, have been intentionally passed over, and if fault be found on this account, lack of room must be the excuse.

The work is for the most part a compilation, and it does not profess to contain much that is original. A word of explanation on this point is necessary, as, except in rare instances, it has not been thought desirable to mention other authors. A consecutive style has been adopted, which would have been broken by frequent quotations or references. The greater portion of the matter which is not given in the older text-books will be found in the *Journal of the Institute of Actuaries*, to which the initials *J.I.A.*, sometimes appearing in the following pages, refer.

A competent knowledge of algebra has been assumed throughout, and no attempt has been made to bring down the demonstrations, or to explain the results, to those who do not possess such knowledge. It is unsafe for the unlearned to deal with subjects which, from the very nature of the case, are beyond their capacity. Great care has, however, been taken to make all the analysis clear and intelligible to an average mathematician; and although some passages, especially perhaps in the third chapter, may not be found easy, it is hoped that, by attention and patience on the part of the reader, all difficulties will be overcome. In many places more fulness of explanation would have been a decided gain had space permitted.

To render the book complete in itself, at least in so far as students are concerned, a few tables have been appended, so that the reader may exercise himself as he proceeds, by setting himself numerical examples. It need hardly be said that theoretical knowledge without practical skill is of little use.

London, *April* 1882.

TABLE OF CONTENTS.

CHAPTER I.—ON INTEREST.

CHAPTER II.—ON ANNUITIES-CERTAIN.

CHAPTER III.—ON VARIABLE ANNUITIES.

CHAPTER IV.—ON LOANS REPAYABLE BY INSTALMENTS.

CHAPTER V.—ON INTEREST TABLES.

TABLES.

THEORY OF FINANCE.

CHAPTER I.

On Interest.

1. CAPITAL is wealth appropriated to reproductive employment, and, in the words of the late J. S. Mill, "the gross profit on capital may be distinguished into three parts, which are respectively the remuneration for risk, for trouble, and for the capital itself, and may be called insurance, wages of superintendence, and interest."

It is with the last of these, INTEREST, that we have at present to do.

2. In civilized communities wealth is measured in money, and therefore in our investigations it is of money that we shall speak; but evidently it is immaterial, in laying down general propositions, whether the standard units are of one actual value or of another, so long as we remain consistent with ourselves, and use the same standard throughout. We shall therefore, as the most convenient course to suit all currencies, treat of *units of value*, without specifying to what currency these units belong.

3. The RATE OF INTEREST is, in scientific investigations, the ratio between the interest earned in one unit of time,—generally a year—and the original sum invested, and it may be represented by the amount of interest earned on one unit of money in one unit of time. In ordinary commercial transactions interest is calculated at the rate *per cent.*, instead of at the rate *per unit;* and therefore to convert the commercial rate into the corresponding rate employed in mathematical researches, we must divide by 100. Thus if 5 per cent. be the rate of interest charged by a banker for an advance, the corresponding mathematical rate will be ·05.

The capital invested is called the PRINCIPAL.

4. It is usual to say that there are two kinds of interest, SIMPLE and COMPOUND. If the interest be calculated on the original capital only, for whatever length of time the loan may have been allowed to remain outstanding with interest unpaid, then it is called *Simple Interest;* but if the interest on each occasion of its becoming due be added to the original debt, and if the interest for each succeeding period be calculated on the original debt so increased by all the previous accumulations of interest, then interest is said to be *Compound.*

It will however be found that the assumption of simple interest leads continually to a *reductio ad absurdum,* which is sufficient evidence that a fallacy somewhere lurks in the supposition. Money, whether received under the name of principal or interest, can always be invested to bear interest, and therefore, from the very nature of the case, simple interest is impossible. It is true that borrower and lender may between themselves agree for only simple interest; but such agreement does not prevent the borrower from investing the interest which is thereby allowed to remain in his hands, and securing interest thereon; and it is because this interest on interest is ignored in the doctrine of simple interest, that the mathematical formulas fail.

5. If at SIMPLE INTEREST—

P=The Principal.

S=The Amount to which that principal will accumulate.

n=The Term, or the number of years the principal is under investment.

i=The Rate of Interest, or the interest on 1 for one year.

Then, since the interest on P for n years is evidently niP, and since S is equal to P increased by its interest, therefore

$$S=P(1+ni) \qquad (1)$$

and, by algebraical transformation,

$$P=\frac{S}{1+ni} \qquad (2)$$

$$n=\frac{S-P}{iP} \qquad (3)$$

$$i=\frac{S-P}{nP} \qquad (4)$$

We have defined P and S as Principal and Amount respectively, but equation (2) exhibits the relationship between them in another light. S is seen to be a *Sum Due* at the end of the period, and

P its *Present Value;* and if we substitute these meanings of the symbols for those already given, all the four equations still hold.

6. The difference between a sum due and its present value is called *Discount;* and if we write D for the discount, we have

$$D = S - P \qquad (5)$$

$$= \frac{niS}{1+ni} \qquad (6)$$

$$= niP \qquad (7)$$

When the sum due is unity and the period one year, we write d for D.

7. We have not made any supposition as to the value of n. If n be fractional, and equal to $\frac{1}{m}$, we have the interest on 1 for the mth part of a year equal to $\frac{i}{m}$. For fractional values of n, as well as for integral, equations (1) to (7) hold.

8. Passing to COMPOUND INTEREST, let

P = The Principal; or the Present Value.

S = The Amount to which that principal will accumulate; or the Sum Due.

n = The Term.

i = The Rate of Interest.

v = The Present value of 1 due a year hence.

d = The Rate of Discount; or the discount on 1 for one year.

$i^{(m)}$ = The effective rate of interest when the nominal rate is convertible m times a year.

$\bar{i}$ = The effective rate of interest when the nominal rate is convertible momently.

δ = The Force of Discount; or the Force of Interest.

9. Since i is the interest on 1 for a year, the amount of 1 in a year will be $(1+i)$. But any other Principal, P, will increase in the same proportion, and the amount of P in a year will be $P(1+i)$. The amount of 1 at the end of the first year being $(1+i)$, if that amount be invested, its amount at the end of the second year will be $(1+i)(1+i)$, or $(1+i)^2$. The amount of $(1+i)^2$ again at the end of the third year will be $(1+i)^3$, which is the amount of 1 in three years; and generally the amount of 1 at the end of n years will be $(1+i)^n$, whence

$$S = P(1+i)^n \qquad (8)$$

By self-evident modifications, from this equation we deduce

$$P=\frac{S}{(1+i)^n} \quad . \quad . \quad . \quad . \quad . \quad . \quad . \quad (9)$$

$$n=\frac{\log S-\log P}{\log (1+i)} \quad . \quad . \quad . \quad . \quad . \quad (10)$$

$$i=\left(\frac{S}{P}\right)^{\frac{1}{n}}-1 \quad . \quad . \quad . \quad . \quad . \quad . \quad (11)$$

10. Since $(1+i)$ is the amount of 1 in a year, it follows that 1 is the present value of $(1+i)$ due at the end of a year, whence the present value of 1 is $\frac{1}{(1+i)}$, that is

$$v=\frac{1}{(1+i)} \quad . \quad . \quad . \quad . \quad . \quad . \quad . \quad . \quad (12)$$

Similarly $(1+i)^n$ being the amount of 1 in n years, 1 is the present value of $(1+i)^n$ due at the end of n years; and $\frac{1}{(1+i)^n}$, or by equation (12), v^n, is the present value of 1 due at the end of n years; whence, S being a sum due and P its present value,

$$P=Sv^n \quad . \quad . \quad . \quad . \quad . \quad . \quad . \quad . \quad (13)$$

Equation (13) is equation (9) in different symbols. It shows us that Principal and Present Value, and Amount and Sum Due, respectively are synonymous, and leads us to the definition that the Present Value is that sum which, invested now, will, at the end of the period, have accumulated exactly to the Sum Due.

11. Because $(1+i)^n$ consists of the unit which was originally invested, together with its accumulations of interest during the n years, therefore the accumulated interest alone on 1 in n years is $\{(1+i)^n-1\}$.

12. We have seen, Art. 9, that when the unit of time is a year, and when i is the interest on 1 in a year, then the amount of 1 in a year is $(1+i)$, and in n years $(1+i)^n$. The same principles hold if we change the unit of time. Suppose, then, we call the mth part of a year the unit of time, and the interest on 1 in such unit $\frac{i}{m}$, so that i becomes the *nominal* annual rate of interest. The amount of 1 in m units of time—that is in a year—will be $\left(1+\frac{i}{m}\right)^m$, and in n years $\left(1+\frac{i}{m}\right)^{mn}$, and the interest actually realised on 1 in a year, or the *effective* rate of interest will be

$$i^{(m)}=\left\{\left(1+\frac{i}{m}\right)^m-1\right\} \quad . \quad . \quad . \quad . \quad . \quad (14)$$

Interest under these circumstances is said to be *convertible m* times in a year.

An illustration of the difference between the effective and the nominal rates of interest, between $i^{(m)}$ and i, is afforded in the case of consols. The nominal rate, i, is 3 per cent., or ·03; but the interest is convertible twice a year, and the effective rate, $i^{(m)}$,—in this case $i^{(2)}$,—is $\{(1{\cdot}015)^2-1\}$, or ·030225.

The higher the rate of interest, and also the more frequently interest is convertible, the greater is the difference between the effective and the nominal rates. Thus, if interest be at the rate of ten per cent., and, consequently, $i = {\cdot}1$, then $i^{(2)} = {\cdot}1025$, $i^{(4)} = {\cdot}103813$, and $i^{(8)} = {\cdot}104486$.

13. In theory there is no limit to the magnitude of m. If m become infinite, and interest be convertible momently, we have still the interest on 1 in a year $\left\{\left(1+\frac{i}{m}\right)^m-1\right\}$. In this case the nominal rate of interest has the special symbol δ assigned to it, and we write, where $\bar{i}$ is the effective rate,

$$\bar{i}=\left\{\left(1+\frac{\delta}{m}\right)^m-1\right\} \quad . \quad . \quad . \quad . \quad (15)$$

But by the theory of logarithms, when m increases without limit $\left(1+\frac{x}{m}\right)^m$ has e^x for its limit, where e is the base of the Napierian system of logarithms, and is equal to 2·7182818. . . . We therefore have

$$\bar{i}=(e^\delta-1) \quad . \quad . \quad . \quad . \quad . \quad . \quad . \quad . \quad (16)$$

$$\text{and } \delta=\log_e(1+\bar{i})=\frac{\log_{10}(1+\bar{i})}{{\cdot}4342945} \quad . \quad . \quad . \quad . \quad (17)$$

From this point of view δ is called the *Force of Interest*, and is the nominal yearly rate of interest when the effective rate is $\bar{i}$.

14. The following Tables A and B afford numerical illustration of the difference between the nominal and the effective rates of interest. Table A shows for various nominal rates the corresponding effective rates when interest is convertible half-yearly, quarterly, or momently, and, conversely, Table B gives for the effective rates the corresponding nominal rates.

TABLE A.

Nominal Rate.	Effective Rate. Interest convertible		
	Half-Yearly.	Quarterly.	Momently.
·03	·030225	·030339	·030454
·035	·035306	·035462	·035620
·04	·040400	·040604	·040811
·045	·045506	·045765	·046028
·05	·050625	·050945	·051271
·06	·060900	·061364	·061837
·07	·071225	·071859	·072508
·08	·081600	·082432	·083287
·09	·092025	·093083	·094174
·1	·102500	·103813	·105171

TABLE B.

Effective Rate.	Nominal Rate. Interest convertible		
	Half-Yearly.	Quarterly.	Momently.
·03	·029778	·029668	·029559
·035	·034698	·034552	·034401
·04	·039608	·039412	·039221
·045	·044504	·044260	·044017
·05	·049390	·049088	·048790
·06	·059126	·058696	·058269
·07	·068816	·068232	·067659
·08	·078462	·077708	·076961
·09	·088062	·087112	·086178
·1	·097618	·096456	·095310

15. It should be observed that in Arts. 9, 10, and 11, we have used the symbol i in a general sense, for the total interest earned on 1 in a year, and that we have not restricted it to the special case where interest is convertible yearly. From the definitions of the symbols it is an identity that $\left(1+\frac{i}{m}\right)^{mn}=(1+i^{(m)})^n$, and unless it be desired to specially emphasize the fact that interest is convertible in some particular manner, the affix (m) may be omitted, and i may

be taken to represent the interest actually realized at the end of a year by the investment of 1. Looked at broadly from this point of view, all the equations in Arts. 9, 10, and 11 remain true: but so far we have discussed only integral values of n.

16. If i be the interest on 1 for one year, what is the interest for the mth part of a year? This question at one time created a great deal of warm controversy among actuaries; see *J. I. A.* vols. 3 and 4. Some writers maintained that the answer must be $\frac{i}{m}$, and the chief argument they urged in support of their view was that $\frac{i}{m}$ correctly represents the interest in the mth part of a year on the supposition of simple interest, and that under no circumstances should compound interest yield less than simple.

On the other hand, rival authorities asserted that the correct expression is $\{(1+i)^{\frac{1}{m}}-1\}$, and that although this gives a value smaller than $\frac{i}{m}$, it is only right that it should do so. The interest is not due till the end of the year, and if the lender receive it sooner he must be content with less, because, compound interest being supposed, he can invest his interest for the remaining portion of the year and realize interest thereon. Also, the same authorities submitted that $\frac{i}{m}$ is palpably wrong, because if 1 at the end of the mth part of a year amount to $\left(1+\frac{i}{m}\right)$, it must at the end of two such parts amount to $\left(1+\frac{i}{m}\right)^2$, and at the end of a year to $\left(1+\frac{i}{m}\right)^m$, or to $\left(1+i+\frac{m-1}{2m}i^2+\text{etc.}\right)$, which is contrary to the fundamental axiom that i is the interest on 1 for a year. They therefore advocated the principle that in theoretical investigations the equation $S=P(1+i)^n$ must be held to be universally true whether n be integral or fractional; and in recent years the majority of mathematicians have adopted their view.

If $\frac{i}{m}$ be considered to be the fractional interest, great complications must sometimes be introduced into formulas. See for instance the *Treatise on Annuities* by the late Griffith Davies, page 81, and Milne on *Life Annuities*, pp. 13 to 15, where not only are the expressions very complex, but three cases must be studied in the solution of one problem. If on the contrary we adopt $\{(1+i)^{\frac{1}{m}}-1\}$, these difficulties vanish. Our expressions become elegant and compact, and only one case presents itself for examination instead of three. See Baily's *Doctrine of Interest and Annuities*, articles 78 and 79, where the same example is discussed as we have already referred to in the works of Davies and Milne.

It must be remembered, however, that in commercial transactions the interest on 1 for the mth part of a year is taken as $\frac{i}{m}$; and even in actuarial formulas it is sometimes found convenient for purposes of numerical calculation to do the same. The effect of so doing is to omit the powers of i above the first in the expansion of $(1+i)^{\frac{1}{m}}$; and, as the quantity i is always very small, it is evident that there is but little practical difference between the two ways of dealing with fractional interest. That way should be adopted which may be the more convenient for the purposes in hand.

17. By definition, the Discount is the difference between a sum due and its present value; that is

$$d=1-v \qquad (18)$$

$$=\frac{i}{1+i} \qquad (19)$$

$$=vi \qquad (20)$$

18. Equation (20) shows us that the discount is the value at the beginning of a period of the interest to be received at the end; but this fact could have been ascertained by reasoning from first principles, without the aid of algebraical transformations. By Art. 10, 1 is the present value of $(1+i)$ to be received at the end of a year; that is, 1 is the present value of the original unit, together with the present value of the interest upon it; and therefore the difference between the unit and its present value is equal to the present value of the interest. Looked at from this point of view then, the discount is the interest paid in advance. This result is of considerable importance in connection with life annuities and premiums.

Equation (20) also tells us that d is a year's interest on v.

19. From equation (18) we immediately deduce

$$v=1-d \qquad (21)$$

$$1+i=\frac{1}{1-d} \qquad (22)$$

Therefore if the rate of *discount* as distinguished from the rate of *interest* be named, we find the present value of a unit due at the end of a year by subtracting from the unit the rate of discount.

20. In this connection it should be noticed that, when a merchant seeks to discount a bill, his banker quotes to him the rate of *discount*, not the rate of *interest;* and when the Bank of England Directors fix their rate, it is the rate of *discount* they determine, not the rate of *interest.* Through failure to keep this distinction in view, confusion has sometimes arisen. It has been usual to say that commercial discount differs from theoretical discount, in that when the rate is, say, 5 per cent., the banker deducts 5 from his customer's bill of 100, instead of $\frac{5}{1\cdot05}$, as his critics say he ought to do; and some writers have even insinuated dishonesty on the part of the banker. Baily, for instance—*Doctrine of Interest*, etc., chap. iii.—remarks that the course "is neither correct nor just." But if the banker says that his rate of *discount* is 5 per cent., the merchant cannot grumble. The banker merely assigns a value to d from which i may be found. It is true that, at the same nominal rate, money improves faster under the operation of discount than of interest. If a banker can employ his funds in discounting at 5 per cent., it will not be to his profit to grant advances at 5 per cent. interest; and if a merchant sell his bill to the banker at 5 per cent. discount, he must remember that he is paying more than 5 per cent. for the accommodation, but with his eyes open to this fact he suffers no wrong.

21. It is not often in business transactions that discount has to be calculated for more than a year. In fact, the great majority of commercial bills have only a fractional part of a year to run. If that fraction be denoted by $\frac{1}{m}$, and if d be the yearly rate of discount, it is customary for the banker to give for each unit of the bill $\left(1-\frac{d}{m}\right)$, and not $(1-d)^{\frac{1}{m}}$, thus using that which by analogy may be called "simple discount." If "simple discount" be employed for periods greater than a year, erroneous and anomalous results are produced. Thus, if the bill to be discounted have n years to run, its value at simple discount will be $(1-nd)$, and it may very well happen that nd is greater than unity, and gives the bill a negative value. This is not because "commercial discount" differs from "theoretical discount," but because we have used "simple discount," and to simple discount the objections mentioned in Article 4 apply equally as to simple interest. The

correct formula is $(1-d)^n$, by the use of which the anomaly disappears.

22. The operation of discount is similar in its results to the operation of interest, and just as we have "compound interest," we may have, by repeating discount operations, "compound discount." When a bill matures, the banker may at once employ the proceeds in discounting a new bill. If d be the nominal yearly rate of discount, and if the process of discounting be repeated m times in a year, the discount on 1 in the mth part of a year will be $\frac{d}{m}$, and the value of 1 due at the end of the mth part of a year will be $\left(1-\frac{d}{m}\right)$. Repeating the operation, we have the present value of 1 due at the end of two m^{th} parts $\left(1-\frac{d}{m}\right)^2$, and the present value of 1 due at the end of a whole year $\left(1-\frac{d}{m}\right)^m$. In this expression we may make m as great as we please; but by the exponential theorem, when m increases without limit, $\left(1-\frac{x}{m}\right)^m$ has e^{-x} for its limit. Also, when discounting is performed momently, d, the nominal rate of discount, is written δ. We therefore have

$$v=e^{-\delta} \quad . \quad . \quad . \quad . \quad . \quad . \quad . \quad . \quad (23)$$

From this point of view δ is called the *Force of Discount.* The term "force of discount" is more commonly employed than "force of interest."

23. By a very simple application of the differential calculus we can form a clear idea of the meaning of the force of discount or the force of interest.

The differential coefficient of a function is the measure of the velocity of change in the function consequent on change in the variable. Now, v^x may be considered as the function of discount, and, taking its differential coefficient, we have $\frac{dv^x}{dx}=v^x \log_e v$. If now we divide by v^x, we have the measure of the velocity of change in the function for each unit of the function, $\frac{1}{v^x}\frac{dv^x}{dx} = \log_e v$ $=-\log_e(1+i)=-\delta$.

In the same way $(1+i)^x$ may be considered as the function of interest, and $\frac{d\,(1+i)^x}{dx}=(1+i)^x \log_e (1+i)$, and $\frac{1}{(1+i)^x}\frac{d\,(1+i)^x}{dx}$ $=\log_e (1+i)=\delta$.

The numerical value is the same, but the sign is opposite. The difference in sign shows that the force of discount is a force of decrement, and the force of interest a force of increment. We can define that force as the annual rate per unit at which a sum is increasing by interest or diminishing by discount at any moment of time.

The term "force" is a misnomer; it should rather be "velocity." But "force" having come into general use, a change would be inconvenient.

24. The quantities i, v, d, and δ, are all mutually dependent, and can be conveniently expressed in terms of each other by means of series. Thus, since by equation (22)

$$(1+i)=(1-d)^{-1}$$

$$i=d+d^2+d^3+\text{etc.} \qquad (24)$$

Also, since by equation (16), $i=e^\delta-1$; by the exponential theorem

$$i=\delta+\frac{\delta^2}{\lfloor 2}+\frac{\delta^3}{\lfloor 3}+\text{etc.} \qquad (25)$$

Again, by equation (22), $(1-d)=(1+i)^{-1}$. Therefore

$$d=i-i^2+i^3-\text{etc.} \qquad (26)$$

and also $d=1-v$

$$=1-e^{-\delta}$$

$$=-\left\{-\delta+\frac{\delta^2}{\lfloor 2}-\frac{\delta^3}{\lfloor 3}+\text{etc.}\right\}$$

$$=\delta-\frac{\delta^2}{\lfloor 2}+\frac{\delta^3}{\lfloor 3}-\text{etc.} \qquad (27)$$

Since $v=1-d$

$$v=1-i+i^2-i^3+\text{etc.} \qquad (28)$$

and also $v=1-\delta+\frac{\delta^2}{\lfloor 2}-\frac{\delta^3}{\lfloor 3}+\text{etc.} \qquad (29)$

Finally, because $\delta=\log_e(1+i)$, therefore, by the theory of logarithms

$$\delta=i-\frac{i^2}{2}+\frac{i^3}{3}-\text{etc.} \qquad (30)$$

and since $\delta=-\log_e v=-\log_e(1-d)$, therefore

$$\delta=d+\frac{d^2}{2}+\frac{d^3}{3}+\text{etc.} \qquad (31)$$

All the series given in this article are rapidly convergent, and when the numerical value of one of the functions is given, they offer great facilities for computing the values of the others. The student will find it an excellent exercise to set himself examples under this head.

The following Table C gives the values of the several quantities at the rates of interest most commonly in use.

TABLE C.

i.	v.	d.	δ.
·02	·980392	·019608	·019803
·025	·975610	·024390	·024692
·03	·970874	·029126	·029559
·035	·966184	·033816	·034401
·04	·961538	·038462	·039221
·045	·956938	·043062	·044017
·05	·952381	·047619	·048790
·06	·943396	·056604	·058269
·07	·934579	·065421	·067659
·08	·925926	·074074	·076961
·09	·917431	·082569	·086178
·1	·909091	·090909	·095310

25. In how many years will money double itself at compound interest? The answer to this question follows directly from equation (10). Writing 2 for S and 1 for P, we have, using Napierian logarithms, $n=\frac{\log_e 2}{\log_e(1+i)}$. But $\log_e(1+i)=i-\frac{i^2}{2}+\frac{i^3}{3}-$ etc., and if we neglect the second and higher powers of i, and write for $\log_e 2$ its near value ·69, we have approximately $n=\frac{·69}{i}$. Whence the common rule:—To find the number of years in which money will double itself, divide 69 by the rate of interest per cent.

Mr. G. F. Hardy (*Insurance Record*, March 31, 1882) has pointed out that the correction to the approximation given by this rule is very nearly a constant quantity: that is, the error involved is practically the same whatever the rate of interest. Thus

$$\frac{\log_e 2}{\log_e(1+i)}=\frac{·693}{i-\frac{1}{2}i^2+\frac{1}{3}i^3-\text{etc.}}=\frac{·693}{i}(1+\tfrac{1}{2}i-\tfrac{1}{12}i^2+\text{etc.})=\frac{·693}{i}+·35$$

very nearly. With this correction the rule will give a result true usually to two places of decimals.

26. It frequently happens that there are various sums of money due at different times from one merchant to another, and which it is desired to pay all at one time. That time at which they may

be paid without injustice to either party is called the equated time of payment, or the average due date.

27. If the various sums be S_1, S_2, S_3, etc., and the times at which they respectively fall due n_1, n_2, n_3, etc., to find the equated time, x, at which the total amount $(S_1+S_2+S_3+\text{etc.})$ may be paid, so that the parties may be on an equality as regards interest.

It is evident that justice will be secured if the present value of the aggregate of the sums due at the time, x, be equal to the aggregate of the present values of the individual sums, and that to find x we have the equation

$$\frac{S_1+S_2+S_3+\text{etc.}}{(1+i)^x}=\frac{S_1}{(1+i)^{n_1}}+\frac{S_2}{(1+i)^{n_2}}+\frac{S_3}{(1+i)^{n_3}}+\text{etc.} \quad (32)$$

which may be symbolically written

$$\frac{\Sigma S}{(1+i)^x}=\Sigma\frac{S}{(1+i)^n} \quad . \quad . \quad . \quad . \quad . \quad . \quad (33)$$

If we expand each side of equation (32) by the binomial theorem and neglect all powers of i above the first, we have, after reduction, and division by the coefficient of x,

$$x=\frac{S_1n_1+S_2n_2+S_3n_3+\text{etc.}}{S_1+S_2+S_3+\text{etc.}} \quad . \quad . \quad . \quad . \quad (34)$$

28. The last equation gives us the usual rule, which is approximately correct:—To find the equated time of payment of various amounts, multiply each amount by the time to elapse until it will fall due, and divide the sum of the products by the sum of the amounts.

29. If simple interest be assumed, the subject becomes intricate, and, as explained in Art. 4, yields contradictory results if regarded from different points of view. The reader may refer to Todhunter's *Algebra*, chapter "Equation of Payments."

CHAPTER II.

On Annuities-Certain.

1. AN Annuity is a periodical payment, lasting during a fixed term of years, or depending on the continuance of a given life or combination of lives. The annuity may be payable either yearly or at more frequent intervals, but it is measured by the total amount payable in one year, which is sometimes conveniently called the annual rent. Thus if a person be entitled to receive £25 every three months, he is said to be in possession of an annuity of £100, payable by quarterly instalments. When an annuity is to last during a fixed term of years it is called an annuity-certain. When it depends on the continuance of a given life or combination of lives, it is called a life annuity, or simply an annuity.

2. The word *status* is conveniently used to denote the period during which an annuity is payable, whether that period be a fixed term of years, or depend for its duration on the contingencies of life. We may therefore define an annuity, generally, as a periodical payment lasting during a given status.

3. A perpetuity is an annuity that is to last for ever. Such are the dividends on consols.

4. Unless otherwise stated the first payment of an annuity is supposed to be made at the end of the first year for which the annuity is to run, or, in the case of annuities payable at shorter intervals, at the end of the first interval. Thus, if we speak of an annuity for n years, we mean an annuity consisting of n yearly payments, the first of which is to be made a year hence: or if we speak of an annuity for n years deferred m years, we mean an annuity consisting of n yearly payments, the first of which is to be made at the end of $(m+1)$ years. This last annuity is said to be

entered on at the end of m years, although the first payment will not be made until the expiration of $(m+1)$ years.

If the first payment of the annuity be made at the beginning instead of at the end of the first interval, the annuity is called an *annuity-due.*

5. For purposes of investigation we shall always consider the annual rent of an annuity to be 1. Our results will be made available for other annuities by merely multiplying by the annual rent.

6. Let

$s_{\overline{n}|}$ = the amount of an annuity for n years.

$a_{\overline{n}|}$ = the present value of an annuity for n years payable yearly.

$a^{(2)}_{\overline{n}|}$ = do. half-yearly.

$a^{(4)}_{\overline{n}|}$ = do. quarterly.

$a^{(m)}_{\overline{n}|}$ = do. m times in a year.

$\bar{a}_{\overline{n}|}$ = do. momently, that is, a continuous annuity.

$\text{a}_{\overline{n}|}$ = the present value of an annuity-due for n years so that $\text{a}_{\overline{n}|} = 1 + a_{\overline{n-1}|}$.

${}_{m|}a_{\overline{n}|}$ = the present value of an annuity for n years, deferred m years.

a_{∞} = the present value of a perpetuity.

NOTE.—The symbols $\text{a}_{\overline{n}|}$, ${}_{m|}a_{\overline{n}|}$, and a_{∞}, may be qualified by the affixes (2), (4), (m), in the same way as an ordinary annuity. We may also write ${}_{m|}\bar{a}_{\overline{n}|}$ and $\bar{a}_{\infty}$.

At Simple Interest.

7. To find $s_{\overline{n}|}$

When the payments of an annuity are not taken as they fall due, but are allowed to remain to accumulate at interest, the annuity is sometimes said to be *forborne*, and the sum to which the payments accumulate is called the *amount* of the annuity. Thus the amount of an annuity is the sum of the amounts of its several payments. Therefore, the last payment having just been made, the last but one having been made a year ago, and so on, we have, commencing with the last payment,

$$s_{\overline{n}|} = 1 + (1+i) + (1+2i) + \ldots + (1+\overline{n-1}i)$$

$$= \frac{n}{2}\left\{2 + \overline{n-1}i\right\} \quad . \quad . \quad . \quad . \quad . \quad (1)$$

8. To find $a_{\overline{n}|}$.

The present value of an annuity is the sum of the present values of its several payments. We therefore have, beginning with the first payment,

$$a_{\overline{n}|}=\frac{1}{1+i}+\frac{1}{1+2i}+\frac{1}{1+3i}+\ \ldots\ +\frac{1}{1+ni}\quad . \qquad (2)$$

There is no direct means of summing this series, and to obtain exactitude, each term must be calculated separately. A very close approximation may however be obtained from the formula of the differential and integral calculus given in Art. 10.

9. Several other so-called approximations have been proposed, based on plausible reasoning which however fails when simple interest is assumed. Some of these are as follows:—

α. It is evident that the present value of an annuity and the present value of the amount of the same annuity should be equal, for, as regards present value, it should be a matter of indifference whether a person is to receive the annuity during the n years, or the equivalent of the annuity at the end of the n years: therefore $a_{\overline{n}|}=\frac{s_{\overline{n}|}}{1+ni}$, and, substituting for $s_{\overline{n}|}$ its value found above,

$$a_{\overline{n}|}=\frac{n\{2+(n-1)i\}}{2(1+ni)}\quad . \quad . \quad . \quad . \quad . \qquad (3)$$

β. Again, the value of the annuity should be equal to the difference between the values of a perpetuity to be entered on at once and another to be entered on at the end of n years. But 1 invested now will yield i at the end of each year for ever: that is, 1 is the value of a perpetuity of i, and therefore the value of a perpetuity of 1 is $\frac{1}{i}$, and the value of a perpetuity of 1 deferred n years is $\frac{1}{(1+ni)i}$, whence

$$a_{\overline{n}|}=\frac{1}{i}\left(1-\frac{1}{1+ni}\right)\quad . \quad . \quad . \quad . \quad . \quad . \qquad (4)$$

γ. Again, from the nature of the case, each payment of the annuity must contain interest on the purchase money, together with a return of part of that purchase money. The purchase money being $a_{\overline{n}|}$ the interest on it is $ia_{\overline{n}|}$, and the return of capital is the balance of the yearly payment, namely $(1-ia_{\overline{n}|})$. This last quantity constitutes the rent of an annuity which must be accumulated at interest in order to replace the capital at the end of the period. That is

$$a_{\overline{n}|}=s_{\overline{n}|}\ (1-ia_{\overline{n}|}):\ \text{or}$$

$$a_{\overline{n}|}=\frac{s_{\overline{n}|}}{1+is_{\overline{n}|}}\quad . \quad . \quad . \quad . \qquad (5)$$

10. The formula of the calculus referred to in Art. 8 is as follows:—

$$\Sigma_0^n V = \int_0^n V dx + \frac{1}{2}(V_0 + V_n) - \frac{1}{12}\left\{\left(\frac{dV}{dx}\right)_0 - \left(\frac{dV}{dx}\right)_n\right\}$$
$$+ \frac{1}{720}\left\{\left(\frac{d^3V}{dx^3}\right)_0 - \left(\frac{d^3V}{dx^3}\right)_n\right\} - \text{etc.}$$

where V is any function of x and Σ_0^n, its finite integral. It is skilfully used by Mr. Woolhouse, *J. I. A.*, vol. xv. p. 100, in the solution of problems in life contingencies, and there he also demonstrates the formula. In the present case $V_x = \frac{1}{1+xi}$, and the formula becomes

$$a_{\overline{n}|} = \frac{\log_e(1+ni)}{i} - \frac{1}{2}\left(1 - \frac{1}{1+ni}\right)$$
$$+ \frac{i}{12}\left(1 - \frac{1}{(1+ni)^2}\right) - \frac{i^3}{120}\left(1 - \frac{1}{(1+ni)^4}\right) \quad . \quad (6)$$

Remembering that the modulus M of the common system of logarithms is ·4342945, and that $\log_e(1+ni) = \frac{\log_{10}(1+ni)}{M}$, the formula is very easily applied.

11. As an illustration of the foregoing formulas, we give, at 5 per cent. interest, the values they bring out for an annuity for twenty years.

$a_{\overline{20}|}$ = 13·616068, by direct summation (true value).
= 14·750000, by formula (3).
= 10·000000, ,, ,, (4).
= 11·919192, ,, ,, (5).
= 13·616068, ,, ,, (6).

It will be seen that formula (6) gives the result correct to six places of decimals, but that the others are very wide of the mark.

Annuities at simple interest have no practical importance, and the analysis is inserted here purely as a matter of curiosity.

Passing to COMPOUND INTEREST

12. To find $s_{\overline{n}|}$

The amounts of the several payments form a geometrical progression, with common ratio $(1+i)$, and we have, beginning as before with the last payment of the annuity,

$$s_{\overline{n}|} = 1 + (1+i) + (1+i)^2 + \quad . \quad . \quad . \quad + (1+i)^{n-1}$$
$$= \frac{(1+i)^n - 1}{i} \quad . \quad . \quad . \quad . \quad . \quad . \quad (7)$$

13. This result can be easily obtained without having recourse to series. A unit invested produces an annuity of i per annum. The amount of 1 in n years consists of the original unit and the

amount of the annuity of i for n years which the unit produced; and therefore the amount of the annuity of i is $\{(1+i)^n-1\}$, and, by simple proportion, the amount of an annuity of 1 is $\frac{(1+i)^n-1}{i}$ as before.

14. If the annuity be payable p times in a year, and the interest be convertible q times, to find the amount of the annuity. The amount of 1 in a year is $\left(1+\frac{i}{q}\right)^q$, (Chapter I., Art. 12), and in the pth part of a year $\left(1+\frac{i}{q}\right)^{\frac{q}{p}}$, (Chapter I., Art. 16). The amount of the annuity, beginning with the last payment, is therefore a geometrical progression of pn terms, with common ratio $\left(1+\frac{i}{q}\right)^{\frac{q}{p}}$, and we have, each payment of the annuity being $\frac{1}{p}$,

$$s=\frac{1}{p}\left\{1+\left(1+\frac{i}{q}\right)^{\frac{q}{p}}+\left(1+\frac{i}{q}\right)^{\frac{2q}{p}}+\ .\ .\ .\ +\left(1+\frac{i}{q}\right)^{\frac{(np-1)q}{p}}\right\}$$

$$=\frac{1}{p}\frac{\left(1+\frac{i}{q}\right)^{nq}-1}{\left(1+\frac{i}{q}\right)^{\frac{q}{p}}-1} \qquad (8)$$

when $q=p$ this becomes

$$s=\frac{\left(1+\frac{i}{q}\right)^{nq}-1}{i} \qquad (9)$$

15. We see that in all cases the amount of an annuity is the total interest on 1 during the whole currency of the annuity, divided by the product of the number of instalments per annum into the interest on 1 during the interval between two instalments of the annuity.

16. To find $a_{\overline{n}|}$

The present value of the annuity consists of a geometrical progression of n terms, the first term and the common ratio both being v. We therefore have

$$a_{\overline{n}|}=v+v^2+v^3+\text{ etc. }+v^n$$

$$=v\frac{1-v^n}{1-v}$$

$$=\frac{1-v^n}{i} \qquad (10)$$

17. By reasoning similar to that in Art. 13 we can arrive at equation (10).

A unit paid down is the value of an annuity of i for n years (produced by the investment of the unit for the term), and of the original unit to be returned at the end of the term: that is $(1-v^n)$ is the value of an annuity of i for n years, and consequently the value of an annuity of 1 is $\frac{1-v^n}{i}$.

18. Also, by means of perpetuities, we can obtain the same result.

A unit invested will yield an annuity of i for ever. Therefore the value of a perpetuity of 1 is

$$a_\infty = \frac{1}{i} \quad . \quad . \quad . \quad . \quad . \quad (11)$$

Now, an annuity for n years is a perpetuity entered on at once, less a perpetuity deferred n years, and its value is $a_\infty\,(1-v^n)$ or $\frac{1-v^n}{i}$.

19. If the annuity be payable p times in a year, and interest be convertible q times, we have

$$a = \frac{1}{p}\left\{\left(1+\frac{i}{q}\right)^{-\frac{q}{p}} + \left(1+\frac{i}{q}\right)^{-\frac{2q}{p}} + \text{etc.} + \left(1+\frac{i}{q}\right)^{-nq}\right\}$$

$$= \frac{1}{p}\,\frac{1-\left(1+\frac{i}{q}\right)^{-nq}}{\left(1+\frac{i}{q}\right)^{\frac{q}{p}}-1} \quad . \quad (12)$$

and when $p=q$ this becomes

$$a = \frac{1-\left(1+\frac{i}{q}\right)^{-nq}}{i} \quad . \quad (13)$$

20. In all cases the value of an annuity is the total discount on 1 during the whole currency of the annuity, divided by the product of the number of instalments per annum and the interest on 1 during the interval between two instalments of the annuity.

21. If the intervals between the payments of the annuity become indefinitely small, so that the payments are made momently, the annuity is said to be continuous.

22. To find the amount and the present value of a continuous annuity, interest convertible momently, we must in formulas (9) and (13) make q infinite, when i becomes δ. We then have, agreeably with Chapter I. Art. 13,

$$\bar{s}_{\overline{n}|} = \frac{e^{n\delta}-1}{\delta} \quad (14)$$

$$\bar{a}_{\overline{n}|} = \frac{1-e^{-n\delta}}{\delta} \quad . \quad . \quad . \quad . \quad . \quad . \quad . \quad (15)$$

23. If, however, the annuity be not continuous, but be payable p times in a year, while interest is convertible momently, we must in formulas (8) and (12) make q infinite, when i becomes δ. We then obtain

$$s=\frac{1}{p}\,\frac{e^{n\delta}-1}{e^{\frac{\delta}{p}}-1} \qquad (16)$$

$$a=\frac{1}{p}\,\frac{1-e^{-n\delta}}{e^{\frac{\delta}{p}}-1} \qquad (17).$$

When the annuity is payable yearly these become

$$s=\frac{e^{n\delta}-1}{e^{\delta}-1} \qquad (18)$$

$$a=\frac{1-e^{-n\delta}}{e^{\delta}-1} \qquad (19)$$

24. Again, if the annuity be continuous while interest is convertible q times a year, p becomes infinite in the formulas

$$s=\frac{1}{p}\,\frac{\left(1+\frac{i}{q}\right)^{nq}-1}{\left(1+\frac{i}{q}\right)^{\frac{q}{p}}-1}$$

$$a=\frac{1}{p}\,\frac{1-\left(1+\frac{i}{q}\right)^{-nq}}{\left(1+\frac{i}{q}\right)^{\frac{q}{p}}-1}$$

In the denominators, p being infinite, there appears the indeterminate form $p\left\{\left(1+\frac{i}{q}\right)^{\frac{q}{p}}-1\right\}$, the value of which is not at once evident, but it may be ascertained easily as follows :—

$$p\left\{\left(1+\frac{i}{q}\right)^{\frac{q}{p}}-1\right\}=p\left\{\left[1+\frac{q}{p}\cdot\frac{i}{q}+\frac{\frac{q}{p}\left(\frac{q}{p}-1\right)}{\underline{|2}}\left(\frac{i}{q}\right)^2+\text{etc.}\right]-1\right\}$$

$$=q\left\{\frac{i}{q}+\frac{\frac{q}{p}-1}{\underline{|2}}\left(\frac{i}{q}\right)^2+\frac{\left(\frac{q}{p}-1\right)\left(\frac{q}{p}-2\right)}{\underline{|3}}\left(\frac{i}{q}\right)^3+\text{etc.}\right\}$$

Now, when p is infinite, $\frac{q}{p}=0$, and the above becomes

$$q\left\{\frac{i}{q}-\frac{1}{2}\left(\frac{i}{q}\right)^2+\frac{1}{3}\left(\frac{i}{q}\right)^3-\text{ etc. }\right\}$$

the value of which, by the theory of logarithms, is, $q\log_e\left(1+\frac{i}{q}\right)$, or $\log_e\left(1+\frac{i}{q}\right)^q$

We then finally have, where the annuity is payable continuously, and interest is convertible q times a year

$$s=\frac{\left(1+\frac{i}{q}\right)^{nq}-1}{\log_e\left(1+\frac{i}{q}\right)^q} \quad . \quad . \quad . \quad .. \quad (20)$$

$$a=\frac{1-\left(1+\frac{i}{q}\right)^{-nq}}{\log_e\left(1+\frac{i}{q}\right)^q} \quad . \quad (21)$$

25. The following two tables illustrate the preceding articles. Table A gives algebraical expressions for the value of an annuity payable yearly, half-yearly, quarterly, m times a year, and momently, with interest convertible at like intervals; and Table B gives corresponding numerical values where the term is twenty-five years and the rate of interest is 4 per cent.

Table A applies also to perpetuities. For a perpetuity, the algebraical term of the numerator will in each case vanish. Thus the value of a yearly perpetuity, interest convertible m times a year, is $\dfrac{1}{\left(1+\frac{i}{m}\right)^m-1}$

The student will find it a useful exercise to re-calculate the second table for himself. Where logarithms to the base e appear in the formulas, the calculations should be made as if all the logarithms were to the base 10, and the final result should be multiplied by the modulus ·434294482.

26. It will be seen from Table A that if the times at which interest is convertible remain unchanged, we can find from the value of an annuity payable yearly the value of one payable at any other intervals, by multiplying by a quantity which is constant for all values of n. Thus if i represent the *effective* rate of interest, no matter how often it may be convertible, we have $a_{\overline{n}|}=\dfrac{1-(1+i)^{-n}}{i}$,

and $a^{(m)}_{\overline{n}|}=\dfrac{1-(1+i)^{-n}}{m\{(1+i)^{\frac{1}{m}}-1\}}$: whence $a^{(m)}_{\overline{n}|}=a_{\overline{n}|}\times\dfrac{i}{m\{(1+i)^{\frac{1}{m}}-1\}}$

where the co-efficient of $a_{\overline{n}|}$ is independent of the term of the annuity.

It should be noted that $m\{(1+i)^{\frac{1}{m}}-1\}$ is the nominal rate of interest convertible m times a year when the effective rate is i. If therefore we write j for that nominal rate, we have $a^{(m)}_{\overline{n}|}=a_{\overline{n}|}\times\dfrac{i}{j}$.

Also $\bar{a}_{\overline{n}|}=a_{\overline{n}|}\times\dfrac{i}{\delta}$.

TABLE A.—VALUE OF ANNUITY OF 1 FOR n YEARS.

ANNUITY PAYABLE	INTEREST CONVERTIBLE Yearly.	Half-yearly.	Quarterly.	m times a year.	Momently.
Yearly, ...	$\frac{1-(1+i)^{-n}}{i}$	$\frac{1-\left(1+\frac{i}{2}\right)^{-2n}}{\left(1+\frac{i}{2}\right)^{2}-1}$	$\frac{1-\left(1+\frac{i}{4}\right)^{-4n}}{\left(1+\frac{i}{4}\right)^{4}-1}$	$\frac{1-\left(1+\frac{i}{m}\right)^{-mn}}{\left(1+\frac{i}{m}\right)^{m}-1}$	$\frac{1-e^{-n\delta}}{e^{\delta}-1}$
Half-Yearly, . .	$\frac{1}{2}\,\frac{1-(1+i)^{-n}}{(1+i)^{\frac{1}{2}}-1}$	$\frac{1-\left(1+\frac{i}{2}\right)^{-2n}}{i}$	$\frac{1}{2}\,\frac{1-\left(1+\frac{i}{4}\right)^{-4n}}{\left(1+\frac{i}{4}\right)^{2}-1}$	$\frac{1}{2}\,\frac{1-\left(1+\frac{i}{m}\right)^{-mn}}{\left(1+\frac{i}{m}\right)^{\frac{m}{2}}-1}$	$\frac{1}{2}\,\frac{1-e^{-n\delta}}{e^{\frac{\delta}{2}}-1}$
Quarterly, . . .	$\frac{1}{4}\,\frac{1-(1+i)^{-n}}{(1+i)^{\frac{1}{4}}-1}$	$\frac{1}{4}\,\frac{1-\left(1+\frac{i}{2}\right)^{-2n}}{\left(1+\frac{i}{2}\right)^{\frac{1}{2}}-1}$	$\frac{1-\left(1+\frac{i}{4}\right)^{-4n}}{i}$	$\frac{1}{4}\,\frac{1-\left(1+\frac{i}{m}\right)^{-mn}}{\left(1+\frac{i}{m}\right)^{\frac{m}{4}}-1}$	$\frac{1}{4}\,\frac{1-e^{-n\delta}}{e^{\frac{\delta}{4}}-1}$
m times a year, .	$\frac{1}{m}\,\frac{1-(1+i)^{-n}}{(1+i)^{\frac{1}{m}}-1}$	$\frac{1}{m}\,\frac{1-\left(1+\frac{i}{2}\right)^{-2n}}{\left(1+\frac{i}{2}\right)^{\frac{2}{m}}-1}$	$\frac{1}{m}\,\frac{1-\left(1+\frac{i}{4}\right)^{-4n}}{\left(1+\frac{i}{4}\right)^{\frac{4}{m}}-1}$	$\frac{1-\left(1+\frac{i}{m}\right)^{-mn}}{i}$	$\frac{1}{m}\,\frac{1-e^{-n\delta}}{e^{\frac{\delta}{m}}-1}$
Momently, . . .	$\frac{1-(1+i)^{-n}}{\delta}$	$\frac{1-\left(1+\frac{i}{2}\right)^{-2n}}{\log_e\left(1+\frac{i}{2}\right)^{2}}$	$\frac{1-\left(1+\frac{i}{4}\right)^{-4n}}{\log_e\left(1+\frac{i}{4}\right)^{4}}$	$\frac{1-\left(1+\frac{i}{m}\right)^{-mn}}{\log_e\left(1+\frac{i}{m}\right)^{m}}$	$\frac{1-e^{-n\delta}}{\delta}$

N.B.—The symbol δ in the above Table has not quite the same meaning in the first as in the last column. In the first column i is the effective rate, and therefore $\delta=\log_e(1+i)$. In the last column i has become the nominal rate, and therefore i itself is written δ. Thus in the first column of Table B, δ is $\log_e 1{\cdot}04$, or $\cdot 039221$, while in the last column, δ is $\cdot 04$.

TABLE B.

VALUE OF ANNUITY FOR 25 YEARS AT 4 PER CENT.

ANNUITY PAYABLE	INTEREST CONVERTIBLE			
	Yearly.	Half-yearly.	Quarterly.	Momently.
Yearly	15·62208	15·55624	15·52282	15·48906
Half-yearly	15·77677	15·71180	15·67883	15·64551
Quarterly	15·85449	15·78998	15·75722	15·72413
Momently	15·93236	15·86840	15·83588	15·80301

27. From equation (15) by making n infinite we have the value of a continuous perpetuity,

$$\bar{a}_{\infty} = \frac{1}{\delta} \quad . \quad . \quad . \quad (22)$$

and similarly from equation (19) we have at momently interest the value of a perpetuity payable yearly,

$$a_{\infty} = \frac{1}{i} \quad . \quad . \quad . \quad . \quad . \quad (23)$$

We might have derived these results from Table A as explained in Art. 25.

28. We have seen that when n is integral, the value of $a_{\overline{n}|}$—that is the value of an annuity for n years—is $\frac{1-v^n}{i}$. By analogy $a_{\overline{n+\frac{1}{m}}|} = \frac{1-v^{n+\frac{1}{m}}}{i}$, and we must investigate the meaning of this expression. We have

$$a_{\overline{n+\frac{1}{m}}|} = \frac{1-v^{n+\frac{1}{m}}}{i} = \frac{1-v^n}{i} + \frac{(1-v^{\frac{1}{m}})v^n}{i}$$

$$= a_{\overline{n}|} + \frac{\{(1+i)^{\frac{1}{m}}-1\}}{i} v^{n+\frac{1}{m}} \quad . \quad . \quad . \quad (24)$$

It therefore appears that $a_{\overline{n+\frac{1}{m}}|}$, the annuity in question, is equal to $a_{\overline{n}|}$, the ordinary annuity for n years, together with a sum of $\frac{\{(1+i)^{\frac{1}{m}}-1\}}{i}$ to be paid at the end of $\left(n+\frac{1}{m}\right)$ years. But $\{(1+i)^{\frac{1}{m}}-1\}$ is on the supposition of compound interest (Chapter I. Art. 16), the interest on 1 for the mth part of a year, or, in other words, the proportion for the mth part of a year of an annual payment of i; and therefore $\frac{\{(1+i)^{\frac{1}{m}}-1\}}{i}$ is the proportion for the mth part of a year of an annual payment of 1. Therefore $a_{\overline{n+\frac{1}{m}}|}$ represents, on the supposition of compound interest throughout, the value of an annuity for $\left(n+\frac{1}{m}\right)$ years. Smart tabulates the values of this expression for annuities for fractional parts of a year, and he has been much criticized for so doing. We have shown, however, that—theoretically—the course he has followed is the correct one. In practice, when an annuity is to run for $\left(n+\frac{1}{m}\right)$ years, it is customary to make the last payment $\frac{1}{m}$, and not $\frac{\{(1+i)^{\frac{1}{m}}-1\}}{i}$.

29. The present value of an annuity for n years deferred t years is evidently

$$
\begin{aligned}
{}_{t|}a_{\overline{n}|} &= v^{t+1}+v^{t+2}+v^{t+3}+\ \ldots\ +v^{t+n} \\
&= v^t\frac{1-v^n}{i} \\
&= v^t a_{\overline{n}|},\ \text{or}\ = a_{\overline{t+n}|}-a_t \qquad (25
\end{aligned}
$$

This formula affords us the means of determining the fine which should be paid for the renewal of the expired term of a lease. Thus, if there be a lease which was originally granted for n years, but of which only t years remain, and if it be desired to renew the expired term at the same rent as before, while there has been an increase of K per annum in the value of the property, the tenant must pay the landlord the value of an annuity of K per annum for $(n-t)$ years—the term for which the lease is to be renewed—deferred t years,—the term for which the present lease has still to run. The value of such an annuity by formula (25) is $Kv^t a_{\overline{n-t}|}$, or $K(a_{\overline{n}|}-a_{\overline{t}|})$.

If the lease had been originally granted for a sum paid down, without the reservation of an annual rent, then of course in determining the fine, we must take account of the whole present annual value of the property, and not only of the increase of that value.

30. If a lease be granted for t years, with the proviso that at the end of that time the tenant shall have the option of renewing it for a like term on payment of a fine f, and so on for ever, we shall have (writing F for the value of all future fines) F equal to $f(v^t+v^{2t}+v^{3t}+\text{ etc.})$; that is,

$$F=f\frac{v^t}{1-v^t}=f\frac{1}{(1+i)^t-1}$$

In fact, the fines constitute a perpetuity in which the unit of time is t years, and consequently at rate of interest $\{(1+i)^t-1\}$.

The foregoing formula assumes that the next fine will be payable t years hence. If the next fine be payable at once, we shall have $F=f\frac{(1+i)^t}{(1+i)^t-1}$, or $=f\frac{1}{1-v^t}$; and generally, if the next fine be payable m years hence, then

$$F=f\frac{v^m}{1-v^t} \qquad (26)$$

The same principles apply where the number of future fines is limited.

It will be noticed that fines for the renewal of leases are only rent under another name, paid down in lump sums instead of year by year.

31. We have already remarked, in investigating formula (5), that each payment of an annuity must contain interest on the purchase-money, together with a return of part of that purchase-money. The buyer of an annuity expects to receive interest on his investment at the agreed rate, and he would not be content unless, at the end of the term, his capital were still intact. In fact, the rent of the annuity consists of two portions; first, interest on the purchase-money; and, secondly, a repayment of capital, called the *sinking fund;* and while the interest may be treated as income by the annuitant, he must scrupulously set aside annually the sinking fund, and reserve it, with all accumulations of interest upon it at the agreed rate, in order to replace his capital at the end of the period when his annuity will expire.

The capital invested being $a_{\overline{n}|}$, the interest on it is $ia_{\overline{n}|}$, and the sinking fund is the balance of the annual rent, or $(1-ia_{\overline{n}|})$.

By the conditions of the case, the sinking fund accumulated must amount to the capital, or in symbols $a_{\overline{n}|}=(1-ia_{\overline{n}|})\,s_{\overline{n}|}$, or

$$a_{\overline{n}|}=\frac{s_{\overline{n}|}}{1+i\,s_{\overline{n}|}} \qquad . \quad . \quad . \quad . \quad . \quad (27)$$

It is thus seen that the sinking fund is of the nature of an annuity which must accumulate at interest till it amounts to the original capital.

32. But we may look at the matter in another way. In the last article we have considered the whole advance to remain outstanding during all the currency of the annuity, and the sinking fund to be separately invested, to accumulate so as suddenly to extinguish the debt at the end of the period. We may now imagine each portion of capital in the successive payments of the annuity to be at once applied towards liquidating the debt, which will thus gradually diminish until it finally vanishes. As the debt is being paid off, a less and less proportion of the annuity will be required for interest, and a greater and greater proportion will be available to refund the capital.

In reality, however, the two ways of viewing the transaction are the same. In Art. 31 we have supposed the sinking fund to be invested in separate securities till it amounts to the debt, while in the present article we have practically supposed the sinking fund to be invested in the security of the debt itself.

33. To separate the successive payments of an annuity into their component parts of principal and interest. The value of the annuity, or, in other words, the capital invested, is $\frac{1-v^n}{i}$, which is the amount unpaid at the beginning of the first year. During the first year the debt will increase by the operation of interest to $(1+i)\,\frac{1-v^n}{i}$, and at the end of the year a payment of 1 will be made. The amount outstanding at the end of the first year will therefore be $\left\{(1+i)\,\frac{1-v^n}{i}-1\right\}$:—and similarly for any other year. Thus, generally, if to the amount unpaid at the end of any year, just after that year's rent of the annuity has been paid, we add one year's interest and deduct 1, the remainder will be the capital unpaid at the end of the next year. We therefore have the outstanding capital at the end of each year as follows :—

YEAR.	CAPITAL OUTSTANDING.
First, . .	$\left\{(1+i)\dfrac{1-v^n}{i}-1\right\}$, or $\dfrac{1-v^{n-1}}{i}$
Second, .	$\left\{(1+i)\dfrac{1-v^{n-1}}{i}-1\right\}$, or $\dfrac{1-v^{n-2}}{i}$
Third, . . .	$\left\{(1+i)\dfrac{1-v^{n-2}}{i}-1\right\}$, or $\dfrac{1-v^{n-3}}{i}$
etc.	etc.

and, taking the general term, we have the capital unpaid at the end of the mth year.

$$\left\{(1+i)\frac{1-v^{n-m+1}}{i}-1\right\}, \text{ or } \frac{1-v^{n-m}}{i} \quad . \quad . \quad (28)$$

We therefore see that the capital outstanding at the end of the mth year is the value of an annuity for the remainder of the period, $(n-m)$ years, and this is evidently as it ought to be.

34. By the last article, the capital outstanding at the end of $(m-1)$ years is $\dfrac{1-v^{n-m+1}}{i}$. A year's interest thereon is $(1-v^{n-m+1})$, which is the interest included in the mth payment of the annuity. The capital included in the same payment must therefore be v^{n-m+1}.

The capital contained in the mth payment of the annuity being v^{n-m+1}, and in the $(m+1)^{th}$, v^{n-m}, or $(1+i)v^{n-m+1}$, we see that the successive instalments of capital form a geometrical progression with common ratio $(1+i)$. This is as it should be, for if we represent by C_m the capital contained in the mth payment, then that amount of capital being paid off, the interest on it, iC_m, is no longer required, and may be applied, along with C_m, at the end of next year to liquidate the debt, when the amount available for that purpose will be C_m+iC_m, or $C_m(1+i)$ as before. Thus $C_{m+1}=C_m(1+i)$.

35. The redemption of a loan by means of a sinking fund is frequently spoken of as the amortisation of the loan.

36. It very often happens that loans are granted in consideration of a terminable annuity. Municipal corporations borrow in this way on the security of the rates, and limited owners of real property are empowered by certain Acts of Parliament to raise money in exchange for a rent-charge for the purpose of improving the estate. In order to escape the income-tax which would otherwise be demanded on the whole annual payment, it is usual to insert in the deed creating the charge a schedule showing the amount of capital and interest respectively contained in each payment of the annuity.

37. If K be the sum to be advanced, the equivalent rent-charge for n years will be $\frac{K}{a_{\overline{n}|}}$. The sinking fund, that is the capital to be repaid at the end of the first year, will be $K\left(\frac{1}{a_{\overline{n}|}}-i\right)$, or $\frac{K}{s_{\overline{n}|}}$, and, in accordance with Art. 34, the sinking fund multiplied continuously by $(1+i)$ will give the successive repayments of capital. In preparing the schedule, the accuracy of our work may be periodically checked, say at every tenth value, by means of the principles laid down in Art. 33. The annual rent-charge being $\frac{K}{a_{\overline{n}|}}$, the capital contained in the mth payment will be $\frac{v^{n-m+1}K}{a_{\overline{n}|}}$, which is our verification formula. If each tenth value be correct, the intermediate values must be correct too, because our calculations are done by a continuous process. Logarithms will materially expedite the work, as, starting with log $\frac{K}{s_{\overline{n}|}}$, we shall merely have to add continuously $\log(1+i)$.

If the rent-charge is to be paid half-yearly, the same principles apply. We can consider the annuity to be for $2n$ years at rate of interest $\frac{i}{2}$.

38. From the last article it may be gathered that

$$\frac{1}{a_{\overline{n}|}}-i=\frac{1}{s_{\overline{n}|}} \quad . \quad . \quad . \quad . \quad . \quad . \quad . \quad (29)$$

that is, the annuity for n years which a sum will purchase, less a year's interest on the purchase money, is equal to the sinking fund which will accumulate to the sum in the n years. In interest tables it is therefore unnecessary to display the values of both $\frac{1}{a_{\overline{n}|}}$ and $\frac{1}{s_{\overline{n}|}}$, as, if one of these quantities be given, the other can be immediately found by inspection.

39. A numerical example will enable us to understand more clearly the principles laid down in the preceding articles. A sum of £5000 is to be advanced at 5 per cent. interest, and to be repaid in fifteen years by equal annual instalments, including principal and interest. It is required to separate each year's payment into principal and interest.

Here, by Art. 37, the yearly charge is $\frac{5000}{a_{\overline{15}|}}$, or £481·711, and the sinking fund is $(481{\cdot}711 - i \times 5000)$ or $\frac{5000}{s_{\overline{15}|}}$, or £231·711. In the following schedule the transaction is worked out at length. The amount of principal opposite any year in column 3 is obtained by multiplying by $(1+i)$, or 1·05, the amount opposite the previous year, and it will further be found that the amount opposite any year, the mth, is equal to $481{\cdot}711v^{n-m+1}$, or $481{\cdot}711v^{16-m}$. Thus, for the 10th year, $v^6 = {\cdot}7462154$, and $v^6 \times 481{\cdot}711 = 359{\cdot}460$.

The principal repaid to date given in column 4 is the sum up to date of column 3. It will also be found to be the amount of an annuity of the sinking fund for the period elapsed. Thus, the amount of an annuity for 10 years is 12·578, and the sinking fund being as above 231·711, the product is 2914·440, which is also the amount opposite year 10 in column 4 of the table.

Also the principal outstanding as given in column 5 is obtained by deducting all the previous repayments of principal as given in column 4 from the original £5000; but also that opposite the mth year is equal to $a_{\overline{n-m}|} \times 481{\cdot}711$. Thus, opposite year 10, the amount is 2085·560, which is 5000−2914·440. But $a_{\overline{n-m}|}$ or $a_{\overline{5}|}$ is 4·3295 and $a_{\overline{5}|} \times 481{\cdot}711 = 2085{\cdot}560$ as before.

SCHEDULE illustrating the Redemption of £5000 in 15 years by Equal Annual Instalments of £481·711, including Principal, and Interest at 5 per cent.

1. Year (m).	2. Interest contained in m^{th} payment.	3. Principal contained in m^{th} payment.	4. Principal repaid to date.	5. Principal still outstanding.
1	250·000	231·711	231·711	4768·289
2	238·414	243·297	475·008	4524·992
3	226·249	255·462	730·470	4269·530
4	213·476	268·235	998·705	4001·295
5	200·064	281·647	1280·352	3719·648
6	185·982	295·729	1576·081	3423·919
7	171·196	310·515	1886·596	3113·404
8	155·670	326·041	2212·637	2787·363
9	139·368	342·343	2554·980	2445·020
10	122·251	359·460	2914·440	2085·560
11	104·277	377·434	3291·874	1708·126
12	85·406	396·305	3688·179	1311·821
13	65·590	416·121	4104·300	895·700
14	44·784	436·927	4541·227	458·773
15	22·938	458·773	5000·000	

40. It sometimes happens, especially in transactions connected with mining property, that, in consideration of a terminable annuity, a lender grants an advance at a higher rate of interest than he can secure from other investments, and that he wishes to realize the higher rate on the whole of the capital during the entire term of the annuity, although, as we have seen, part of his capital is repaid to him year by year. He must therefore so fix the price he pays for the annuity that he may realize interest at the higher rate, called the remunerative rate, and accumulate the sinking fund at a lower rate, called the accumulative rate. To effect this object we must apply the principles of formula (27). We have only to take i at the remunerative rate, and s at the accumulative rate. Writing A for the value of the annuity under these special conditions, and s' for the amount of an annuity at the accumulative rate, we have

$$A=\frac{s'}{1+is'} \quad . \quad . \qquad (30)$$

These terms are very onerous to the borrower, as he has not only to pay the higher rate of interest on the capital which remains in his possession, but also to make up to that higher rate, during the whole period for which the annuity has to run, the interest on the accumulations of the sinking fund which are in the hands of the lender.

41. As an example of the last article, let it be required to ascertain the annual sum to be paid in order to redeem a debt in n years, interest at the rate i to be realised on the whole debt throughout the whole term, and the instalments of capital to be invested at another rate, j.

We have seen that the value of an annuity of 1 per annum to pay the purchaser interest at the rate i on his purchase-money and to refund that purchase-money at another rate, j, is $\frac{s'}{1+is'}$. For each unit of the debt the annual payment is therefore $\frac{1}{A}$, or $\left(\frac{1}{s'}+i\right)$ or $\frac{1}{a'}+(i-j)$, $\left(\text{since } \frac{1}{s'}=\left\{\frac{1}{a'}-j\right\}\right)$, where a' is taken at the rate j.

42. This last formula enables us to use a table of ordinary annuities in questions of this more complex character. If, for instance, we have a table giving the ordinary annuities which 1

will purchase, and we wish to find the annuity which 1 will purchase to pay i to the purchaser and replace the capital at rate j, we have only to take from the table the annuity which 1 will purchase at rate j, and add to it $(i-j)$.

43. A numerical illustration will be useful, and it will be convenient to repeat with the needful modifications the one already given in Art. 39. By this means a comparison can be instituted between the ordinary annuity and the special annuity now under consideration.

Suppose a lender to advance £5000, and we are asked to find the annuity for 15 years which will pay him 7 per cent. on his capital and replace that capital at 5 per cent.

The annuity for 15 years which 1 will purchase at 5 per cent. is ·0963423. Adding to this $(i-j)$ or ·02, and multiplying by 5000, we have £581·711 the annual payment required. This is greater than in the example in Art. 39 by £100, or 2 per cent. on £5000.

The sinking fund is to be taken simply at 5 per cent., and must therefore be the same as in Art. 39, viz. £231·711.

The annual rent of the annuity is, as above, £581·711, which is equal to the sinking fund, together with £350 or 7 per cent. on £5000. The lender may therefore treat the £350 as income during the whole of the 15 years, if he invest annually the sinking fund at 5 per cent. to replace his capital. But—as in Article 32—we may consider that part of the capital is repaid year by year, and that on the repaid portions of the capital the lender only realizes 5 per cent. interest. The annual rent of the annuity must then supply an instalment of capital, interest at 7 per cent. on the capital still outstanding, and also interest at 2 per cent. on the capital already repaid, so as to raise to 7 per cent. the interest on that portion of capital also.

The operation of the fund is shown in the following schedule, by which it is seen that all these requirements are fulfilled. Columns 3, 4, and 5 are identical with the columns bearing the same headings in Art. 39; but the interest in column 6 of the present schedule is always 2 per cent. more than that in the corresponding column, column 2, of the former schedule, while we have added column 8, showing the balance of interest on the principal already repaid. It will be seen that the sum of the amounts opposite any year in columns 3, 6, and 8, is always equal to the annual rent of the annuity, and that the sum of the amounts in columns 6, 7, and 8 is always £350, or 7 per cent. on the total capital advanced.

SCHEDULE showing the operation of an Annuity to repay £5000 in 15 years at 7 per cent. interest, Sinking Fund accumulating at 5 per cent.

1. Year (m).	2. Annual Rent of Annuity.	3. Principal included in m^{th} Payment.	4. Principal repaid to date.	5. Principal still outstanding.	6. Interest at 7 per cent. on outstanding Principal at end of previous year (provided by Annuity).	7. Interest at 5 per cent. earned on repaid Principal at end of previous year.	8. Balance of Interest on repaid Principal at end of previous year, being difference between 7 per cent. and 5 per cent. (provided by Annuity).
1	581·711	231·711	231·711	4768·289	350·000	0·000	0·000
2		243·297	475·008	4524·992	333·780	11·586	4·634
3		255·462	730·470	4269·530	316·749	23·751	9·500
4		268·235	998·705	4001·295	298·867	36·524	14·609
5		281·647	1280·352	3719·648	280·090	49·936	19·974
6		295·729	1576·081	3423·919	260·375	64·018	25·607
7		310·515	1886·596	3113·404	239·674	78·804	31·522
8		326·041	2212·637	2787·363	217·938	94·330	37·732
9		342·343	2554·980	2445·020	195·115	110·632	44·253
10		359·460	2914·440	2085·560	171·151	127·749	51·100
11		377·434	3291·874	1708·126	145·989	145·723	58·288
12		396·305	3688·179	1311·821	119·569	164·594	65·837
13		416·121	4104·300	895·700	91·827	184·410	73·763
14		436·927	4541·227	458·773	62·699	205·216	82·085
15		458·773	5000·000	0·000	32·114	227·062	90·824

44. In connection with annuities, various quantities come under our notice, namely, the term, the present value, the amount, the annual rent, and the rate of interest, and in the majority of cases simple relations subsist between these quantities, so that, certain of them being given, the others can at once be found by the ordinary operations of algebra. Thus we have seen that $a_{\overline{n}|} = \frac{1-(1+i)^{-n}}{i}$, so that, having given n and i, we can find a.

45. There is one case, however, which presents difficulties. When the term, and the amount or the present value, of an annuity are given, and it is required to ascertain the rate of interest, we have to solve an equation of the nth degree, and an approximate solution is alone possible. This case we now proceed to investigate.

46. Having given s, the amount of an annuity, and n, the term, to find i, the rate of interest. Using Table 3, which shows the

amounts of annuities at various rates of interest, we find, in the table opposite the given number of years, the value nearest to s. Let that value be denoted by s', and let the rate of interest under which it is found be denoted by j. If s', the amount found in the table, be exactly equal to s, then the rate under which s' is found must be the rate sought; that is, $i=j$. If, however, s' differ from s, then j differs from i. Let $i=j+\rho$, and the problem resolves itself into finding ρ, which is necessarily a very small quantity.

We have
$$s=\frac{(1+i)^n-1}{i}$$
$$=\frac{\{(1+j)+\rho\}^n-1}{j+\rho}.$$

Therefore $s\,(j+\rho)=\{(1+j)+\rho\}^n-1$.

If we expand the right-hand member of the equation by the binomial theorem, and neglect all powers of ρ above the first, we have
$$s\,(j+\rho)=(1+j)^n+n\,(1+j)^{n-1}\rho-1\,;$$
whence, remembering that $(1+j)^n-1=js'$, we have
$$\rho=\frac{j\,(s'-s)}{s-n\,(1+j)^{n-1}} \quad . \quad . \quad . \quad . \quad . \quad . \quad (31)$$

By means of formula (31) the value of ρ, and thence of i, can be found, generally with sufficient accuracy; but should a closer approximation be desired, the process may be repeated. The amount of the annuity may be calculated at the rate of interest found by the first application of the formula, and inserted in the equation instead of s', by which means a result very accurate indeed may be obtained.

47. Other formulas for the rate of interest in $s_{\overline{n}|}$ may be deduced, analogous to those given in the succeeding articles for the rate of interest in $a_{\overline{n}|}$; but, as the problem is not one of great importance, we do not prolong the investigation here.

48. Having given a, the value of an annuity, and n, the term, to find i, the rate of interest.

By the help of Table 4, giving the values of annuities, we must find a value of a near to the given value. Let that near value be denoted by a', and let the rate of interest under which it is found be denoted by j, and let the rate sought, $i=j+\rho$.

We have
$$a=\frac{1-(1+i)^{-n}}{i}$$
$$=\frac{1-\{(1+j)+\rho\}^{-n}}{j+\rho}$$

whence $a\,(j+\rho)=1-\{(1+j)+\rho\}^{-n}$ . . . . . (32)

and $a=[1-\{(1+j)+\rho\}^{-n}]\,(j+\rho)^{-1}$ . . . (33)

From these last two equations we can deduce as follows the four approximate formulas A, B, C, and D, for ρ.

49. If in equation (32) we expand by the binomial theorem, neglect the powers of ρ above the second, and collect the terms, we have,—writing v for $(1+j)^{-1}$, and remembering that $(1-v^n)=ja'$,—the quadratic equation in ρ,

$$\frac{n(n+1)}{2}v^{n+2}\rho^2+(a-nv^{n+1})\rho-j(a'-a)=0 \quad . \quad (34)$$

If we neglect the second power of ρ, we have merely a simple equation to solve, from which we obtain

$$\text{A.} \qquad \rho=\frac{j(a'-a)}{a-nv^{n+1}} \quad . \quad . \quad . \quad . \quad . \quad (35)$$

50. The value just found in A may be used in conjunction with equation (34) to obtain a closer approximation. If in the equation we write $\rho^2=\rho\times\frac{j(a'-a)}{a-nv^{n+1}}$, and then solve for ρ, we have

$$\text{B.} \qquad \rho=\frac{j(a'-a)}{a-nv^{n+1}+\frac{n(n+1)}{2}\frac{j(a'-a)v^{n+2}}{a-nv^{n+1}}} \quad . \quad (36)$$

51. Returning now to equation (33), if we expand both the factors of the right-hand member by the binomial theorem, and retain only the first and second powers of ρ, we have another equation

$$\frac{1}{j^2}\left\{a'-nv^{n+1}-\frac{n(n+1)}{2}v^{n+2}j\right\}\rho^2-\frac{1}{j}(a'-nv^{n+1})\rho+(a'-a)=0 \quad (37)$$

and if in that equation we solve for ρ, neglecting the second power, we have

$$\text{C.} \qquad \rho=\frac{j(a'-a)}{a'-nv^{n+1}} \quad . \quad . \quad . \quad . \quad . \quad (38)$$

52. Lastly, if in equation (34) we write $\rho^2=\rho\times\frac{j(a'-a)}{a'-nv^{n+1}}$, and solve for ρ, we have

$$\text{D.} \qquad \rho=\frac{j(a'-a)}{a-nv^{n+1}+\frac{n(n+1)}{2}\frac{j(a'-a)v^{n+2}}{a'-nv^{n+1}}} \quad . \quad (39)$$

53. A numerical example will help us to gain a clearer idea of the formulas given in the four preceding articles. Suppose that 14·94390 is the value of an annuity-certain for 30 years, and we require to find the rate of interest.

If in Table 4 we look against 30 years, we find that at 5 per cent. $a_{\overline{30}|}=15{\cdot}37245$, and at 6 per cent., $a_{\overline{30}|}=13{\cdot}76483$. We therefore conclude that i, the rate of interest sought, lies between ·05 and ·06, being nearer to ·05, and we assume that $j=$·05. Then, using formula A,

$$a'=15{\cdot}37245 \qquad \text{also } v^{31}= {\cdot}220359$$
$$a=14{\cdot}94390 \qquad 30v^{31}=6{\cdot}61077$$
$$(a'-a)= {\cdot}42855$$
$$j(a'-a)= {\cdot}021428 \qquad a-30v^{31}=8{\cdot}33312$$
$$\log {\cdot}021428=\bar{2}{\cdot}33098$$
$$\log 8{\cdot}33312 =0{\cdot}92081$$
$$\log \rho=\bar{3}{\cdot}41017$$
$$\rho= {\cdot}0025714$$

Formula B has the appearance of being lengthy and intricate, but it is not so. It should be observed that the second term of the denominator consists merely of the product of $\frac{n(n+1)}{2}v^{n+2}$ and the value of ρ found by formula A, and that in finding the value of ρ by formula A we have already calculated the values of all the other terms of the expression. In the example in hand we have

$$\log (\tfrac{1}{2}\times 30\times 31)=2{\cdot}66745$$
$$\log v^{32} =\bar{1}{\cdot}32194$$
$$\log \frac{j(a'-a)}{a-30v^{31}} =\bar{3}{\cdot}41017$$
$$\bar{1}{\cdot}39956$$

whence the second term of the denominator is ·25093. Adding to this the value of $a-30v^{31}$ already found, 8·33312, we have the denominator=8·58405. The numerator is ·021428 as before, whence

$$\rho={\cdot}0024963.$$

Formula C is the same as A, except that a' occurs instead of a in the denominator. From it

$$\rho={\cdot}0024457.$$

Formula D is the same as B, except that a' occurs instead of a in the second term of the denominator. From it

$$\rho={\cdot}0024998.$$

We therefore have by formula

A. $i={\cdot}0525714$
B. $i={\cdot}0524963$
C. $i={\cdot}0524457$
D. $i={\cdot}0524998$

the true value of i being ·0525.

54. Formula C gives a better value than formula A, and, if we examine the processes by which the formulas were deduced, the reason becomes apparent.

Equation (32) is obtained from equation (33) by multiplying by $(j+\rho)$, with the result of a loss of accuracy. If in the expansion of $1-\{(1+j)+\rho\}^{-n}$ we denote the coefficients of the ascending powers of ρ by k, l, and m, so that $k=\{1-(1+j)^{-n}\}$, $l=n(1+j)^{-(n+1)}$, and $m=\frac{n(n+1)}{2}(1+j)^{-(n+2)}$, equation (32) will take the form

$$m\rho^2+(a-l)\rho-(k-aj)=0,$$

or say

$$x\rho^2+y\rho-z=0 \quad . \quad . \quad . \quad . \quad . \quad . \quad . \quad (40)$$

where $x=m$, $y=(a-l)$, and $z=(k-aj)$, and equation (33) will take the form

$$ja=(k+l\rho-m\rho^2)\left(1-\frac{\rho}{j}+\frac{\rho^2}{j^2}\right)$$

or

$$\left(m+\frac{l}{j}-\frac{k}{j^2}\right)\rho^2+\left(a-l+\frac{k-aj}{j}\right)\rho-(k-aj)=0$$

which is seen to be the same as

$$\left(x+\frac{y}{j}\right)\rho^2+\left(y+\frac{z}{j}\right)\rho-z=0 \quad . \quad . \quad . \quad . \quad . \quad (41)$$

Equations (40) and (41) are simply equations (34) and (37) so displayed that they may be compared, and they show us that in equation (34) we neglect part of each power of ρ. In fact, when, to obtain equation (34), we multiply up by $(j+\rho)$, the result is to diminish the coefficient of each power of ρ by $\frac{1}{j}th$ part of the coefficient of the next lower power.

Seeing then that formulas A and C are identical in form, and therefore equally easily applied, while C is the more accurate of the two, that is the one which should be used.

55. Formula A was first published by Mr. Francis Baily in the appendix, page 129, of his work on the Doctrine of Interest and Annuities. In the same place he investigates other formulas which do not require the aid of interest tables for their application, but at the present day these formulas have no practical importance, owing to the number and extent of the interest tables which are available.

Formula C is due to Mr. George Barrett, who, in writing to Mr. Baily, suggests it as an improvement on formula A. (See the extract of a letter from Barrett to Baily, *J. I. A.* vol. iv. p. 189.)

To Professor De Morgan we owe formula B. It is given by him in a paper, *J. I. A.* vol. viii. p. 67; and Mr. M'Lauchlan has supplied formula D, *J. I. A.* vol. xviii. p. 295.

56. The principles by which we reduced equation (34) from a

quadratic into a simple equation may obviously be extended. We may retain all terms of our original expansion which involve powers of ρ not greater than the third, and write $\rho^2=\rho\left\{\frac{j(a'-a)}{a-nv^{n+1}}\right\}$, and $\rho^3=\rho\left\{\frac{j(a'-a)}{a-nv^{n+1}}\right\}^2$, and then solve for ρ. By this means Professor De Morgan (*J. I. A.* vol. viii. p. 67) deduces a formula which produces very accurate results, but it is too complicated for common use.

57. A very frequent instance of the application of the foregoing formulas is the case of foreign government loans. The English Government has usually borrowed in exchange for perpetual annuities,—that is, in consideration of a sum advanced a promise has been given to pay a certain yearly interest, but without stipulation as to repayment of the principal.

58. Foreign Governments, however, generally find it more convenient to undertake to repay the principal within a limited period. The arrangement is to issue numbered bonds, with coupons attached representing the annual interest, and these coupons are cut off by the lender and presented for payment as the interest falls due. Besides providing for the interest yearly, the Government also sets aside a fixed sum, called the sinking fund, to redeem the bonds; and bonds to the amount of the sinking fund are drawn by lot and cancelled. The yearly interest, however, on the cancelled bonds is still provided by the Government, and applied, in addition to the sinking fund, to liquidate the loan. In fact the Government simply contracts to pay a fixed yearly instalment, including principal and interest, for a limited number of years till the whole loan is paid off, or, in other words, it raises money in exchange for a terminable annuity. The sinking fund under these conditions is called an accumulative sinking fund.

59. If such a rate of interest were offered that the public would take the bonds at par, the transaction would not be complex. The rate of interest paid by the Government would merely be the nominal rate of interest borne by the bonds. But loans of this class generally are issued at a discount, while the bonds on being drawn are paid off at par, so that, in addition to the yearly interest, the lending public are offered the inducement of a bonus on repayment. Under these conditions we must make use of one of our approximate formulas to find the true rate of interest paid by the Government.

60. Let the price at which the bonds are issued be k per unit: let the nominal rate of interest paid by the Government be i'; and let the sinking fund per unit of the loan, that is the amount to be repaid at the end of the first year, be z:—then k is simply the present value of an annuity of $(i'+z)$ per annum. We therefore have $k=(i'+z)\frac{1-(1+i)^{-n}}{i}$

$$\frac{k}{(i'+z)}=\frac{1-(1+i)^{-n}}{i} \quad . \quad . \quad . \quad . \qquad (42)$$

where i is the actual, as distinguished from the nominal, rate of interest paid by the borrower, and n is the number of years for which the loan will run.

61. In the above equation both i and n are unknown quantities, but n is easily found. The sinking fund, being employed to pay off the bonds, accumulates at the nominal rate of interest i', and at the end of n years amounts to the total loan. We therefore have the equation

$$1=z\frac{(1+i')^n-1}{i'}$$

whence

$$\frac{i'}{z}+1=(1+i')^n$$

and $$n=\frac{\log\left(\frac{i'}{z}+1\right)}{\log(1+i')} \quad . \quad . \quad . \quad . \quad . \quad . \quad . \qquad (43)$$

Having found n we can then apply one of our formulas A, B, C, or D, to find i. Usually n will not be an integer, but it will be sufficient for practical purposes if we take the nearest integral value which equation (43) brings out. Instead of using equation (43), it will most often be convenient to refer to an interest table to find the value of n.

62. As an example, let us examine the Russian 1864 loan of £6,000,000. The issue price was 85 per cent., the nominal rate of interest 5 per cent., and the accumulative sinking fund 1 per cent. That is, in equation (42), $k=\cdot85$, $i'=\cdot05$, and $z=\cdot01$; therefore $\frac{k}{i'+z}=14\cdot1667$. Also in equation (43) $n=\frac{\log 6}{\log 1\cdot05}=36\cdot71=37$ nearly: whence $a_{\overline{37}|}=14\cdot1667$. If we apply formula C, using $6\frac{1}{4}$ per cent. for the approximate rate j, we find $i=\cdot063297$.

It need hardly be remarked that we can use interest tables to give an approximate rate even when the tables do not contain minute subdivisions of the rate of interest. For instance, in the present case, if the table employed contain annuities at only integral rates of interest, we shall find that at 5 per cent. $a_{\overline{37}|}=16\cdot71129$, and at 6 per cent. $a_{\overline{37}|}=14\cdot73678$. We see therefore that our required rate must lie between 6 per cent. and 7 per cent., and must be about $6\frac{1}{4}$ per cent., and we then calculate $a_{\overline{37}|}$ and v^{38} at $6\frac{1}{4}$ per cent., and insert the results in our formula C.

63. An interesting case occurs where a loan, redeemable by an accumulative sinking fund, is quoted in the market at other than par some years after issue, and we require to ascertain the rate of interest which it yields. Let us take the following illustration. A Government 5 per cent. loan was issued eight years ago, repayable by an accumulative sinking fund of 1 per cent. The eighth annual payment is just due but not paid, and the market price of the loan is 102 per cent. What rate of interest does it yield? The full period of the loan was, as we have seen by the last article, 37 years, and the annual payment made by the Government is 6 per cent. of the original amount of the loan. The capital at present outstanding is that which was outstanding when the seventh payment was just made, that is, by formula (28), $6\times a_{\overline{30}|}$, or $92\cdot2347$ for each 100 of the original loan (the annuity being taken of course at 5 per cent. interest). The market value of the capital still outstanding is, at 102 per cent., $94\cdot07939$, and this is the value of an annuity of 6 per annum of thirty payments, first payment due at once, that is of $6(1+a_{\overline{29}|})$, whence $a_{\overline{29}|}=14\cdot6799$; and applying formula C with $5\frac{1}{4}$ per cent. for j, we find the rate of interest to be $5\cdot28008$ per cent.

64. We must be careful not to confound the rate of interest incurred by the Government with the rate or rather rates of interest realized by the lenders. If the loan stand at a discount and be repayable at par, it is evident that the holders of the bonds which happen to be drawn early for repayment will realize a higher rate of interest than will the holders of the bonds that happen to remain undrawn till a late period. Thus, in the case of the loan discussed in Art. 62, a £100 bond will cost a purchaser £85, and if the bond happen to be drawn at the end of the first year, the holder will then receive £5 for interest and £100 for principal, that is for an advance of £85 for a year, he will receive in principal and interest £105, and thus realize interest at the rate of more than $23\frac{1}{2}$ per cent. If, however, the bond be not drawn till the end of the thirty-seven years, the holder will realize interest at the rate of only very little more than $5\frac{3}{4}$ per cent.

In Art. 63 we spoke of the rate of interest *yielded* by a loan, but that must be understood to mean the rate yielded to a purchaser provided he take up the whole loan, or at least a sufficiently large part of it to ensure a fair average of bonds drawn. It cannot be held to refer to the rate yielded by any particular bond.

65. It will be convenient to close this chapter with one or two examples.

α. What is the value of an annuity-certain, taking interest at i for the first n years and j thereafter?

Let us assume that the annuity has to run for $(n+m)$ years in all. The value of the annuity for the first n years is evidently $\frac{1-(1+i)^{-n}}{i}$. When n years have expired, the value of the remaining portion of the annuity will, by the conditions of the question, be $\frac{1-(1+j)^{-m}}{j}$. But at the present time this latter portion is deferred n years, during which period interest is at rate i, and its present value is therefore $(1+i)^{-n}\frac{1-(1+j)^{-m}}{j}$. The whole value of the annuity is thus $\frac{1-(1+i)^{-n}}{i}+(1+i)^{-n}\frac{1-(1+j)^{-m}}{j}$.

β. A person holds a lease at a rent of £20 per annum, with the option of renewing it every seven years by paying a fine of £100. What is the equivalent uniform annual rent? Interest 5 per cent. This question is likely to occur to the tenant at one of the periods for renewal, when he is debating in his mind whether or not he should continue his lease. We may therefore assume that one of the fines is just due. By Art. 30 the value of the future fines is $100\times\frac{1}{1-v^7}$, and this we must convert into a perpetual annuity by dividing by a_∞, or by multiplying by i. The annual rent equivalent to the fines is therefore $\frac{100\times i}{1-v^7}$, which we shall find to be £17·282, making the total uniform annual rent £37·282, or £37, 5s. 8d. If the next fine be not due for seven years, we shall find that the total annual rent will be £32, 5s. 8d.

γ. A loan of £P is to be discharged by $(p+q+r)$ annual instalments compounded of principal and interest. The p instalments are to be of £α each, the q of £β, and the r of £γ. What is the rate of interest, i? We can only approximate to this rate of interest, and it will be sufficient if in our approximation we neglect all powers of i above the first.

We have

$$P=\alpha\frac{1-(1+i)^{-p}}{i}+\beta(1+i)^{-p}\frac{1-(1+i)^{-q}}{i}+\gamma(1+i)^{-(p+q)}\frac{1-(1+i)^{-r}}{i}$$

$$=\alpha\left\{p-\frac{p(p+1)}{2}i\right\}+\beta\{1-ip\}\left\{q-\frac{q(q+1)}{2}i\right\}$$
$$+\gamma\{1-i(p+q)\}\left\{r-\frac{r(r+1)}{2}i\right\}$$

If now we multiply out, still neglecting all powers of i above the first, and arrange the terms, we have

$$i=\frac{p\alpha+q\beta+r\gamma-P}{\alpha\frac{p(p+1)}{2}+\beta\left\{\frac{q(q+1)}{2}+pq\right\}+\gamma\left\{\frac{r(r+1)}{2}+r(p+q)\right\}}$$

8. The following very interesting question appeared in the examination paper set to the candidates at the intermediate examination of the Institute of Actuaries held at Christmas 1874:—

A Parochial Union has obtained certain loans upon the under-mentioned terms, and it now desires to consolidate the debts and redeem them by a single terminable annuity running from 31st December next—

1st Jany. 1856,	£5000,	by 60	equal half-yearly payments,		int.	6 °/。
1st July 1870,	3000,	,, 80	do.	do.	,,	5 °/。
1st Jany. 1872,	2000,	,, 60	do.	do.	,,	$4\frac{1}{2}$ °/。

Required, 1st, the terminable equal annuity payable half-yearly for thirty years from 31st December next; and, 2nd, the rate of interest (approximate) returned upon the debt when consolidated as above.

Here it will be proper to assume that the half-yearly instalment of each of the three original annuities which falls due on 31st December 1874 will be paid, and that the consolidation of the debt will take place immediately subsequent to such payment.

The first step is to ascertain the annuities at present payable by the Parochial Union. By Art. 37 we find the half-yearly charge

for the first debt to be $\frac{5000}{a_{\overline{60}|}}$ at 3 °/。, or 180·665.

,, second ,, $\frac{3000}{a_{\overline{80}|}}$ at $2\frac{1}{2}$°/。, or 87·078.

,, third ,, $\frac{2000}{a_{\overline{60}|}}$ at $2\frac{1}{4}$°/。, or 61·071.

Immediately after 31st December 1874,

of 1st	debt	38	payments will have been made,	and	22	will remain.
,, 2nd	,,	9	do.	do.	71	do.
,, 3rd	,,	6	do.	do.	54	do.

There will therefore remain outstanding, by Art. 33,

Of 1st debt,	$180{\cdot}665 \times a_{\overline{22}\rceil}$	at 3 °/₀,	or 2879·240
„ 2nd do.,	$87{\cdot}078 \times a_{\overline{71}\rceil}$	at $2\frac{1}{2}$°/₀,	or 2879·765
„ 3rd do.,	$61{\cdot}071 \times a_{\overline{54}\rceil}$	at $2\frac{1}{4}$°/₀,	or 1897·995
		Total amount outstanding,	7657·000

We must now take each debt separately, and find—at the rate of interest proper to that debt—the half-yearly annuity of sixty payments which the amount outstanding will purchase. This we shall find to be, by Art. 37,

For 1st debt, . .	$\dfrac{2879{\cdot}240}{a_{\overline{60}\rceil}}$	at 3 °/₀,	or 104·035
„ 2nd do., . .	$\dfrac{2879{\cdot}765}{a_{\overline{60}\rceil}}$	at $2\frac{1}{2}$°/₀,	or 93·170
„ 3rd do., . .	$\dfrac{1897{\cdot}995}{a_{\overline{60}\rceil}}$	at $2\frac{1}{4}$°/₀,	or 57·956
		Total consolidated half-yearly payment,	255·161

The Parochial Union has now to pay in each year, for thirty years, £510·322, in order to liquidate a debt of £7657, and we have to find the rate of interest.

Using formula C, Art. 51, and $5\frac{1}{4}$ per cent. for our approximate rate of interest, we have

$$
\begin{array}{rr}
a' = 14{\cdot}9439 \\
a = 15{\cdot}0043 \\
\hline
-{\cdot}0604 \\
525 \\
\hline
302 \\
12 \\
3 \\
\hline
j(a'-a) = -{\cdot}00317
\end{array}
\qquad\qquad
\begin{array}{rl}
a' = & 14{\cdot}9439 \\
30\,v^{31} = & 6{\cdot}1410 \\
\hline
& 8{\cdot}8029 \\
\log {\cdot}00317 = & \bar{3}{\cdot}50106 \\
\log 8{\cdot}8029 = & 0{\cdot}94463 \\
\hline
\log(-\rho) = & \bar{4}{\cdot}55643 \\
\rho = & -{\cdot}00036 \\
i = & 5{\cdot}214 \text{°/₀}
\end{array}
$$

It will be noticed that in solving this question we have treated the annuity as one of £510·332, payable annually, and so found the annual rate of interest. This we may consider to be the nominal annual rate convertible half-yearly. It would have come practically to the same thing had we taken the case as it really stands in the question, viz., an annuity of 60 payments of 255·161

Points : Selection of Lifes

Mortality among infants in Upper Classes

each, and assumed an approximate rate of interest $2\frac{5}{8}$ per cent. We should thus have obtained the half-yearly rate of interest. The course we have adopted is the more convenient as being better adapted to the majority of published interest tables.

Another method of finding the rate of interest in the foregoing question might suggest itself; but though at first sight it seems correct, it nevertheless gives an erroneous result. Of the whole debt of £7657, there is 2879·240 at 6°/ₒ, 2879·765 at 5°/ₒ, and 1897·995 at $4\frac{1}{2}$°/ₒ, and therefore the average for the whole debt is 5·252°/ₒ, a higher rate than that already found. To take the average would be the correct course if the three portions of the debt were repayable in the same proportions at the same times. But this is not so. Those portions of the debt at the higher rates of interest are repaid by the sinking fund more slowly in the earlier years and more rapidly in the later years than those at the lower rates, and thus, as time advances, the average is destroyed. The only correct way to look at the question is to treat the consolidated debt as the present value of an annuity of the total annual sum payable by the Parochial Union.

CHAPTER III.

On Variable Annuities.

1. In the direct process of the calculus of Finite Differences, we form one set of functions from another by the operation called differencing, that is, if we have a series whose terms are u_1, u_2, u_3, etc., we form another series whose terms are (u_2-u_1), (u_3-u_2), etc., and which we represent by Δu_1, Δu_2, etc., the successive terms of the second series being the differences between the successive terms of the first series. The operation of differencing may be repeated, the series of differences being itself differenced.

2. It is evident that this process may be inverted. Instead of forming a series of differences, we may form a series of sums. We may pass from our original series to another series, of the terms of which the terms of the original series are the differences. Thus if we have the series u_1, u_2, u_3, etc., we can form another series, V_1, V_2, V_3, etc., such that $V_2-V_1=u_1$, $V_3-V_2=u_2$, etc., or, putting the same operation in another form, such that $V_2=V_1+u_1$, $V_3=V_2+u_2$, etc. Further, like the process of differencing, the inverse process of summation may be repeated. We may pass to another series from V_1, V_2, V_3, etc., in the same way that we passed to V_1, V_2, V_3, etc., from u_1, u_2, u_3, etc., and so on without limit.

3. The successive orders of the figurate numbers are series connected with each other in the way described in Art. 2, and they are represented in the following scheme :—

Term (m)	1st order.	2nd order.	3rd order.	4th order.	5th order.	6th order.	7th order.	etc.
1	1	0	0	0	0	0	0	etc.
2	1	1	0	0	0	0	0	etc.
3	1	2	1	0	0	0	0	etc.
4	1	3	3	1	0	0	0	etc.
5	1	4	6	4	1	0	0	etc.
6	1	5	10	10	5	1	0	etc.
7	1	6	15	20	15	6	1	etc.
8	1	7	21	35	35	21	7	etc.
9	1	8	28	56	70	56	28	etc.
10	1	9	36	84	126	126	84	etc.
etc.	etc.	etc.	etc.	etc.	etc.	etc.	etc.	etc.

4. The first order is a series of constants, which, for the sake of convenience, we may assume to be unity.

Each term of the second order is the sum of all the *preceding* terms of the first order, so that the mth term of the second order is the sum of the first $(m-1)$ terms of the first order, the value of which is therefore $(m-1)$.

In the same way the mth term of the third order is the sum of the first $(m-1)$ terms of the second order; and generally the mth term of the rth order is the sum of the first $(m-1)$ terms of the $(r-1)^{th}$ order.

5. If by $t_{\overline{m}|\,\overline{r}|}$ we denote the mth term of the rth order, and use Σ_1^m as a symbol of summation of terms from 1 to m inclusive, the fundamental relation between the orders may be represented by the equation

$$t_{\overline{m}|\,\overline{r}|}=\Sigma_1^{m-1}\,t_{\overline{m}|\,\overline{r-1}|} \quad . \quad . \quad . \quad . \quad . \quad . \quad . \qquad (1)$$

6. Since by the last equation, $t_{\overline{m}|\,\overline{r}|}=\Sigma_1^{m-1}\,t_{\overline{m}|\,\overline{r-1}|}$, and $t_{\overline{m+1}|\,\overline{r}|}=\Sigma_1^{m}\,t_{\overline{m}|\,\overline{r-1}|}$, it follows that

$$t_{\overline{m+1}|\,\overline{r}|}=t_{\overline{m}|\,\overline{r}|}+t_{\overline{m}|\,\overline{r-1}|} \quad . \quad . \quad . \quad . \quad . \quad . \qquad (2)$$

and $t_{\overline{m+1}|\,\overline{r}|}-t_{\overline{m}|\,\overline{r}|}=t_{\overline{m}|\,\overline{r-1}|}$ . . . . . . (3)

Equations (2) and (3) display the fundamental relation between the orders in another light from that supplied by equation (1), and show that the principles of construction laid down in Art. 2 have been carried out. The terms of the $(r-1)^{th}$ order are respectively the differences of the corresponding terms of the rth order.

7. From the method of construction of the successive orders it will be seen that always

$$t_{\overline{r}|\,\overline{r}|}=1 \; . \quad . \quad . \quad . \quad . \quad . \quad . \quad . \quad . \qquad (4)$$

and that, of the rth order, all the terms below the rth are equal to zero.

8. Carrying out the analogy of equation (4), we may assume that $t_{\overline{0}|\,\overline{0}|}=1$, and also $t_{\overline{0}|\,\overline{r}|}=0$. These conventional symbols (like the symbol x^0 with which the reader in his algebraical studies has already become familiar), although they have no meaning in themselves, will be found useful in our subsequent investigations, as they will enable us to make perfectly general the formulas which we shall deduce.

9. It is easy to show that the mth term of the rth order is equal to $\dfrac{(m-1)(m-2)\ldots(m-r+1)}{\lfloor r-1}$. Let u_1, u_2, u_3, etc., be the successive terms of the rth order. Then, by the ordinary formula of Finite Differences,

$$u_m = u_1 + (m-1)\Delta u_1 + \frac{(m-1)(m-2)}{\underline{|2}}\Delta^2 u_1 + \text{etc.}$$
$$+ \frac{(m-1)(m-2)\ .\ .\ .\ (m-r+1)}{\underline{|r-1}}\Delta^{r-1}u_1$$

But the first term of each of the orders except the first, is equal to zero, and the first term of the first order is equal to unity; and these first terms constitute the successive orders of differences of the rth order of numbers. That is, u_1 and all its differences, up to and including the $(r-2)^{th}$ difference, vanish, and the $(r-1)^{th}$ difference is equal to unity. Therefore

$$t_{\overline{m}|\,\overline{r}|} = \frac{(m-1)(m-2)\ .\ .\ .\ (m-r+1)}{\underline{|r-1}} \qquad . \quad . \qquad (5)$$

10. The principles so far investigated can now be applied to the solution of questions connected with variable annuities. Let an annuity of the rth order be an annuity the successive payments of which are the corresponding terms of the rth order of figurate numbers. Thus an annuity of the first order will be an annuity whose payments are all unity; an annuity of the second order will be an annuity whose payments are 0, 1, 2, 3, etc.; an annuity of the third order will be an annuity whose payments are 0, 0, 1, 3, 6, etc.; and, generally, an annuity of the rth order will be an annuity whose m^{th} payment is, by formula (5),

$$\frac{(m-1)(m-2)\ .\ .\ .\ (m-r+1)}{\underline{|r-1}}$$

Let the amount of an annuity for n years of the rth order be denoted by $s_{\overline{n}|\,\overline{r}|}$, and the present value by $a_{\overline{n}|\,\overline{r}|}$.

11. To find $s_{\overline{n}|\,\overline{r}|}$, the amount of an annuity for n years of the rth order,

We have
$$s_{\overline{n}|\,\overline{r}|} = t_{\overline{n}|\,\overline{r}|} + (1+i)t_{\overline{n-1}|\,\overline{r}|} + (1+i)^2 t_{\overline{n-2}|\,\overline{r}|} + \text{etc.}$$
$$+ (1+i)^{n-1}t_{\overline{1}|\,\overline{r}|} + (1+i)^n t_{\overline{0}|\,\overline{r}|}$$
$$s_{\overline{n}|\,\overline{r-1}|} = t_{\overline{n}|\,\overline{r-1}|} + (1+i)t_{\overline{n-1}|\,\overline{r-1}|} + (1+i)^2 t_{\overline{n-2}|\,\overline{r-1}|} + \text{etc.}$$
$$+ (1+i)^{n-1}t_{\overline{1}|\,\overline{r-1}|} + (1+i)^n t_{\overline{0}|\,\overline{r-1}|}$$

Adding the two equations together by the help of formula (2), we have

$$s_{\overline{n}|\,\overline{r}|} + s_{\overline{n}|\,\overline{r-1}|}$$
$$= t_{\overline{n+1}|\,\overline{r}|} + (1+i)t_{\overline{n}|\,\overline{r}|} + (1+i)^2 t_{\overline{n-1}|\,\overline{r}|} + \text{etc.} + (1+i)^{n-1}t_{\overline{2}|\,\overline{r}|} + (1+i)^n t_{\overline{1}|\,\overline{r}|}$$
$$= (1+i)s_{\overline{n}|\,\overline{r}|} + t_{\overline{n+1}|\,\overline{r}|} \qquad . \quad . \quad . \quad . \quad . \quad . \quad . \qquad (6)$$

Therefore, after arranging the terms,

$$s_{\overline{n}|\,\overline{r}|} = \frac{s_{\overline{n}|\,\overline{r-1}|} - t_{\overline{n+1}|\,\overline{r}|}}{i} \qquad . \quad . \quad . \quad . \quad . \qquad (7)$$

By means of this formula we can find $s_{\overline{n}|\,\overline{r}|}$ having given $s_{\overline{n}|\,\overline{r-1}|}$.

12. We have seen, Art. 8, that an annuity of the 0th order consists of but one payment, or rather that the annuity is really only a unit in possession at the beginning of the period. We may therefore infer that the amount of such an annuity is $(1+i)^n$, and its present value unity, and we may use the conventional symbols

$$s_{\overline{n}|\,\overline{0}|}=(1+i)^n \quad . \quad . \quad . \quad . \quad . \quad . \quad . \quad (8)$$

$$a_{\overline{n}|\,\overline{0}|}=1 \quad . \quad . \quad . \quad . \quad . \quad . \quad . \quad . \quad (9)$$

13. By formula (7), $s_{\overline{n}|\,\overline{1}|}=\frac{s_{\overline{n}|\,\overline{0}|}-t_{\overline{n+1}|\,\overline{1}|}}{i}$. But $t_{\overline{n+1}|\,\overline{1}|}=1$, and by formula (8), $s_{\overline{n}|\,\overline{0}|}=(1+i)^n$. Therefore $s_{\overline{n}|\,\overline{1}|}=\frac{(1+i)^n-1}{i}$. This result agrees with formula (7) of Chapter II., and shows that we have correctly interpreted the symbol $s_{\overline{n}|\,\overline{0}|}$

14. From formula (7) we have

$$s_{\overline{n}|\,\overline{0}|}=(1+i)^n$$

$$s_{\overline{n}|\,\overline{1}|}=\frac{(1+i)^n-1}{i}$$

$$s_{\overline{n}|\,\overline{2}|}=\frac{s_{\overline{n}|\,\overline{1}|}-n}{i}$$

$$s_{\overline{n}|\,\overline{3}|}=\frac{s_{\overline{n}|\,\overline{2}|}-\frac{n(n-1)}{2}}{i}$$

$$s_{\overline{n}|\,\overline{4}|}=\frac{s_{\overline{n}|\,\overline{3}|}-\frac{n(n-1)(n-2)}{6}}{i}$$

etc. = etc.

And, to take a numerical example, when $n=5$, and $i=\cdot05$

$s_{\overline{5}|\,\overline{0}|}=$ 1·2762815625

·2762815625 × 20

$s_{\overline{5}|\,\overline{1}|}=$ 5·52563125 $\qquad n=5$

·52563125 × 20

$s_{\overline{5}|\,\overline{2}|}=$ 10·5126250 $\qquad \frac{n(n-1)}{2}=10$

·5126250 × 20

$s_{\overline{5}|\,\overline{3}|}=$ 10·252500 $\qquad \frac{n(n-1)(n-2)}{6}=10$

·252500 × 20

$s_{\overline{5}|\,\overline{4}|}=$ 5·05000 $\qquad \frac{n(n-1)(n-2)(n-3)}{24}=5$

·05000 × 20

$s_{\overline{5}|\,\overline{5}|}=$ 1·

The last result proves the correctness of our work, because an annuity for 5 years of the 5th order consists of but one payment of unity, made at the end of the fifth year.

15. To find $a_{\overline{n}|\,\overline{r}|}$, the present value of an annuity for n years of the rth order.

Following the method adopted in Art. 11,

$$a_{\overline{n}|\,\overline{r}|}=v^0t_{\overline{0}|\,\overline{r}|}+vt_{\overline{1}|\,\overline{r}|}+v^2t_{\overline{2}|\,\overline{r}|}+\text{etc.}+v^nt_{\overline{n}|\,\overline{r}|}$$

$$a_{\overline{n}|\,\overline{r-1}|}=v^0t_{\overline{0}|\,\overline{r-1}|}+vt_{\overline{1}|\,\overline{r-1}|}+v^2t_{\overline{2}|\,\overline{r-1}|}+\text{etc.}+v^nt_{\overline{n}|\,\overline{r-1}|}$$

$$a_{\overline{n}|\,\overline{r}|}+a_{\overline{n}|\,\overline{r-1}|}=v^0t_{\overline{1}|\,\overline{r}|}+vt_{\overline{2}|\,\overline{r}|}+v^2t_{\overline{3}|\,\overline{r}|}+\text{etc.}+v^nt_{\overline{n+1}|\,\overline{r}|}$$

$$=(1+i)a_{\overline{n+1}|\,\overline{r}|}$$

$$=(1+i)a_{\overline{n}|\,\overline{r}|}+v^nt_{\overline{n+1}|\,\overline{r}|}$$

Therefore

$$a_{\overline{n}|\,\overline{r-1}|}=ia_{\overline{n}|\,\overline{r}|}+v^nt_{\overline{n+1}|\,\overline{r}|}$$

$$\text{and } a_{\overline{n}|\,\overline{r}|}=\frac{a_{\overline{n}|\,\overline{r-1}|}-v^nt_{\overline{n+1}|\,\overline{r}|}}{i} \quad . \quad . \quad . \quad . \quad (10)$$

16. By formula (10), $a_{\overline{n}|\,\overline{1}|}=\dfrac{a_{\overline{n}|\,\overline{0}|}-v^n}{i}$ since $t_{\overline{n+1}|\,\overline{1}|}=1$. This result agrees with formula (10) of Chapter II. if for $a_{\overline{n}|\,\overline{0}|}$ we write 1 in accordance with formula (9) of the present chapter.

17. From formula (10) we have

$$a_{\overline{n}|\,\overline{0}|}=1$$

$$a_{\overline{n}|\,\overline{1}|}=\frac{1-v^n}{i}$$

$$a_{\overline{n}|\,\overline{2}|}=\frac{a_{\overline{n}|\,\overline{1}|}-nv^n}{i}$$

$$a_{\overline{n}|\,\overline{3}|}=\frac{a_{\overline{n}|\,\overline{2}|}-\dfrac{n(n-1)}{2}v^n}{i}$$

$$a_{\overline{n}|\,\overline{4}|}=\frac{a_{\overline{n}|\,\overline{3}|}-\dfrac{n(n-1)(n-2)}{6}v^n}{i}$$

$$a_{\overline{n}|\,\overline{5}|}=\frac{a_{\overline{n}|\,\overline{4}|}-\dfrac{n(n-1)(n-2)(n-3)}{24}v^n}{i}$$

18. As an illustration, let it be required to determine the values of the first five orders of annuities for 40 years at 5 per cent. interest.

$v^{40}=$ ·14204568   1·

$40v^{40}=$ 5·6818272   ·14204568

$\dfrac{40\times 39}{2}v^{40}=$ 110·795630   ·85795432 × 20

$a_{\overline{40}|\,\overline{1}|}=$ 17·1590864

$$\frac{40\times39\times38}{2\times3}v^{40}=1403\cdot41132$$

$$\frac{40\times39\times38\times37}{2\times3\times4}v^{40}=12981\cdot5547$$

$$
\begin{array}{lrr}
a_{\overline{40}|\,\overline{1}|}= & 17\cdot1590864 & \\
 & 5\cdot6818272 & \\
\hline
 & 11\cdot4772592 & \times 20 \\
a_{\overline{40}|\,\overline{2}|}= & 229\cdot545184 & \\
 & 110\cdot795630 & \\
\hline
 & 118\cdot749554 & \times 20 \\
a_{\overline{40}|\,\overline{3}|}= & 2374\cdot99108 & \\
 & 1403\cdot41132 & \\
\hline
 & 971\cdot57976 & \times 20 \\
a_{\overline{40}|\,\overline{4}|}= & 19431\cdot5952 & \\
 & 12981\cdot5547 & \\
\hline
 & 6450\cdot0405 & \times 20 \\
a_{\overline{40}|\,\overline{5}|}= & 129000\cdot810 &
\end{array}
$$

19. It will be seen that in passing to higher orders, the amounts and values of annuities increase for a while with great rapidity when the term n is at all considerable, and that the result of the division by i at each stage is to diminish accuracy by reducing the number of decimal places. If, therefore, we wish for extreme accuracy in the higher orders, we must commence with a great many decimal places.

20. To find $a_{\overline{\infty}|\,\overline{r}|}$, the value of a perpetuity of the rth order.

We have seen, formula (10), that $a_{\overline{n}|\,\overline{r}|}=\dfrac{a_{\overline{n}|\,\overline{r-1}|}-v^n t_{\overline{n+1}|\,\overline{r}|}}{i}$. If n increase without limit, then $a_{\overline{n}|\,\overline{r-1}|}$ will become $a_{\overline{\infty}|\,\overline{r-1}|}$; v^n will diminish without limit; and $t_{\overline{n+1}|\,\overline{r}|}$ will increase without limit except in the case where $r=1$. When $r=1$ then $v^n t_{\overline{n+1}|\,\overline{r}|}$ will vanish, when n increases without limit; and as $a_{\overline{n}|\,\overline{0}|}=1$ for all values of n, the value of the perpetuity of the first order is $\dfrac{1}{i}$, the same result as that given in Chapter II., formula (11).

21. Taking the general case, where $v^n t_{\overline{n+1}|\,\overline{r}|}=v^n\times\dfrac{n(n-1)\ldots(n-r+2)}{\underline{|r-1}}$, we shall now prove that for all finite values of r, $v^n t_{\overline{n+1}|\,\overline{r}|}$ vanishes when n becomes infinite. For conciseness we may write $v^n\times\dfrac{n(n-1)\;.\;.\;.\;(n-r+2)}{\underline{|r-1}}=N$. When n increases to $(n+1)$, then N changes from $v^n\times\dfrac{n(n-1)\ldots(n-r+2)}{\underline{|r-1}}$, to $v^{n+1}\times\dfrac{(n+1)n(n-1)\ldots(n-r+3)}{\underline{|r-1}}$, which we may write N'. That is, to obtain N' we multiply N by

$\frac{n+1}{n-r+2}\times\frac{1}{1+i}$. The first factor of this expression may be written in the form $\frac{1}{1-\frac{r-1}{n+1}}$, which shows that although it will always be greater than unity, yet it will diminish with the increase of n, and tend to approach unity as a limit. N' will be greater than N as long as $\frac{n+1}{n-r+2}\times\frac{1}{(1+i)}>1$; that is, as long as $\frac{n+1}{n-r+2}>(1+i)$; or, in other words, $N'=N$ when $\frac{n+1}{n-r+2}=(1+i)$; that is, when $n=\frac{(r-2)(1+i)+1}{i}=R$ say, a finite quantity, and when n still increases N begins to diminish. If M represent the value—which is finite—of N when $n=R$, the value of N becomes successively, as n still increases, $v\frac{R+1}{R-r+2}M$, $v^2\frac{(R+1)(R+2)}{(R-r+2)(R-r+3)}M$, etc.; that is, M is continuously multiplied by a series of factors of the form $v\frac{(R+m)}{R-r+(m+1)}$, each of which is less than the preceding one, and less than $v\frac{R+1}{R-r+2}$, which is less than unity. Therefore when n becomes $(R+m)$, N becomes less than $M\left\{v\frac{R+1}{R-r+2}\right\}^m$. But the coefficient of M being less than unity, diminishes as m increases, and may be made as small as we please by sufficiently increasing m. When therefore m, and consequently n, becomes infinite, N vanishes, and we have

$$a_{\overline{\infty}|\,\overline{r}|}=\frac{a_{\overline{\infty}|\,\overline{r-1}|}}{i} \qquad . \quad . \quad . \quad . \quad . \qquad (11)$$

The value of the perpetuity of the rth order is therefore finite when that of the $(r-1)^{th}$ order is finite. But the value of a perpetuity of the first order being finite, so will be that of the second order, and of the third order, etc.; and generally the value of a perpetuity of the rth order will be finite as long as r is finite.

22. From formula (11) we have

$$a_{\overline{\infty}|\,\overline{0}|}=1$$

$$a_{\overline{\infty}|\,\overline{1}|}=\frac{1}{i}$$

$$a_{\overline{\infty}|\,\overline{2}|}=\frac{1}{i^2}$$

$$a_{\overline{\infty}|\,\overline{3}|}=\frac{1}{i^3}$$

and generally

$$a_{\overline{\infty}|\,\overline{r}|} = \frac{1}{i^r} \qquad (12)$$

23. The reasoning of Art. 21 may appear to the student to be somewhat laboured. That is only because we have avoided the use of the Differential Calculus. If we call in the aid of the Differential Calculus, we can very easily prove that $v^n t_{\overline{n+1}|\,\overline{r}|}$ vanishes when n becomes infinite. The term which we are investigating is $\frac{n(n-1)\ .\ .\ .\ (n-r+2)}{1\,.\,2\,.\,3\ .\ .\ .\ (r-1)}(1+i)^{-n}$, the numerator and denominator of which both increase without limit with the increase of n. To determine the limit of the value of such a fraction, we must substitute for the numerator and denominator their respective differential coefficients, repeating the process, if necessary, until a fraction is obtained of a form exhibiting its ultimate or limiting value. In the case in hand we shall find that the $(r-1)^{th}$ differential coefficients will answer our purpose. Substituting these differential coefficients for the original numerator and denominator, we have $\frac{1}{\{\log_e(1+i)\}^{r-1}(1+1)^n}$, which evidently vanishes when n becomes infinite, and we therefore conclude that $v^n t_{\overline{n+1}|\,\overline{r}|}$ also vanishes.

24. In Art. 17 of Chapter II., we found the value of an annuity of the first order by means of reasoning without having recourse to series, and we may now extend the reasoning to annuities of higher orders.

The mth payment of an annuity of the $(r-1)^{th}$ order, if it be invested immediately on its receipt, will produce at the end of each succeeding year the sum of $i \times t_{\overline{m}|\,\overline{r-1}|}$, and it will itself still remain in hand at the end of the n years for which the original annuity was granted. If, then, each of the payments of the original annuity be thus forborne, at the end of the first year there will be entered on an annuity, arising from interest, for $(n-1)$ years of $i \times t_{\overline{1}|\,\overline{r-1}|}$, at the end of the second year another annuity for $(n-2)$ years of $i \times t_{\overline{2}|\,\overline{r-1}|}$, and so on, each of these annuities, which we may call secondary annuities, running concurrently with all those that have preceded it. Thus, at the end of the mth year the payments of all these secondary annuities will aggregate $i\,(t_{\overline{1}|\,\overline{r-1}|} + t_{\overline{2}|\,\overline{r-1}|} + \text{etc.} + t_{\overline{m-1}|\,\overline{r-1}|})$, which by formula (1) is equal to $i \times t_{\overline{m}|\,\overline{r}|}$. We therefore see that an annuity of the $(r-1)^{th}$ order forborne, will produce for n years an annuity of i of the rth order, and in addition there will of course remain in hand at the end of the n years the aggregate of all the

n payments of the original annuity, the aggregate of these payments being, by formula (1), $t_{\overline{n+1}|\,\overline{r}|}$, and the present value of the aggregate being $v^n t_{\overline{n+1}|\,\overline{r}|}$. Thus we reach the result, $a_{\overline{n}|\,\overline{r-1}|} = i a_{\overline{n}|\,\overline{r}|} + v^n t_{\overline{n+1}|\,\overline{r}|}$; whence, as in formula (10),

$$a_{\overline{n}|\,\overline{r}|} = \frac{a_{\overline{n}|\,\overline{r-1}|} - v^n t_{\overline{n+1}|\,\overline{r}|}}{i}$$

Similarly for perpetuities. If the payments of a perpetuity of the $(r-1)^{th}$ order be forborne, each of them will produce another perpetuity, the mth payment producing a perpetuity of $i\, t_{\overline{m}|\,\overline{r-1}|}$, and the aggregate payment of these secondary perpetuities to be received at the end of the mth year will be

$$i\,(t_{\overline{1}|\,\overline{r-1}|} + t_{\overline{2}|\,\overline{r-1}|} + \text{etc.} + t_{\overline{m-1}|\,\overline{r-1}|}),$$

or by formula (1), $i\, t_{\overline{m}|\,\overline{r}|}$. Therefore $a_{\overline{\infty}|\,\overline{r-1}|} = i\, a_{\overline{\infty}|\,\overline{r}|}$; whence, as in formula (11), $a_{\overline{\infty}|\,\overline{r}|} = \dfrac{a_{\overline{\infty}|\,\overline{r-1}|}}{i}$.

It will be noticed that in the argument in this article we have escaped the difficulty which occupied us in Arts. 21 and 23.

25. To find the amount and the present value of an annuity whose successive payments are u_1, u_2, u_3, etc.

If for u_2, u_3, etc., we substitute their equivalents in terms of u_1 and its differences, we have

$$u_1 = u_1$$
$$u_2 = u_1 + \Delta u_1$$
$$u_3 = u_1 + 2\Delta u_1 + \Delta^2 u_1$$
$$u_4 = u_1 + 3\Delta u_1 + 3\Delta^2 u_1 + \Delta^3 u_1\,;$$

and generally

$$u_m = u_1 + (m-1)\,\Delta u_1 + \frac{(m-1)\,(m-2)}{\underline{|2}}\,\Delta^2 u_1 + \text{etc.}$$

Therefore the given annuity is equivalent to an annuity of u_1 of the first order, together with an annuity of Δu_1 of the second order, and an annuity of $\Delta^2 u_1$ of the third order, etc.

Let s and a be the amount and the present value respectively of the annuity u_1, u_2, u_3, etc., and let the series u_1, u_2, u_3, etc., have $(r-1)$ orders of differences: then

$$s = s_{\overline{n}|\,\overline{1}|}\, u_1 + s_{\overline{n}|\,\overline{2}|}\, \Delta u_1 + s_{\overline{n}|\,\overline{3}|}\, \Delta^2 u_1 + \text{etc.} + s_{\overline{n}|\,\overline{r}|}\, \Delta^{r-1} u_1 \quad (13)$$

$$a = a_{\overline{n}|\,\overline{1}|}\, u_1 + a_{\overline{n}|\,\overline{2}|}\, \Delta u_1 + a_{\overline{n}|\,\overline{3}|}\, \Delta^2 u_1 + \text{etc.} + a_{\overline{n}|\,\overline{r}|}\, \Delta^{r-1} u_1 \quad (14)$$

If the given annuity be a perpetuity, then

$$a = a_{\overline{\infty}|\,\overline{1}|}\, u_1 + a_{\overline{\infty}|\,\overline{2}|}\, \Delta u_1 + \text{etc.} + a_{\overline{\infty}|\,\overline{r}|}\, \Delta^{r-1} u_1$$

$$= \frac{u_1}{i} + \frac{\Delta u_1}{i^2} + \ldots + \frac{\Delta^{r-1} u_1}{i^r} \quad . \quad . \quad . \quad . \quad . \quad (15)$$

26. The great power of the formulas demonstrated in the last article may best be illustrated by some examples.

α. Let it be required to find the value of an annuity certain for n years whose several payments are 1, 2, 3, etc., n. Here $u_1=1$, $\Delta u_1=1$, and all the higher orders of differences of u_1 vanish. Therefore $a=a_{\overline{n}|\,\overline{1}|}+a_{\overline{n}|\,\overline{2}|}$

$$=\frac{(1+i)\,a_{\overline{n}|\,\overline{1}|}-nv^n}{i}$$

β. Similarly, if it be required to find the amount of the same annuity,

$$s=\frac{(1+i)\,s_{\overline{n}|\,\overline{1}|}-n}{i}$$

γ. If the annuity be a perpetuity increasing 1 per annum for ever,

$$a=\frac{1}{i}+\frac{1}{i^2}$$

δ. Let it be required to find the value of an annuity for forty years, whose several payments are 4, 7, 12, 19, etc.; interest 5 per cent. Here $u_1=4$, $\Delta u_1=3$, and $\Delta^2 u_1=2$. Therefore

$$a=4a_{\overline{40}|\,\overline{1}|}+3a_{\overline{40}|\,\overline{2}|}+2a_{\overline{40}|\,\overline{3}|}$$

Taking the values of the annuities as given in Art. 18, $a=5507{\cdot}253$.

ε. Find the value of an annuity for fifteen years whose several payments are 21, 39, 54, 66, etc. Here $u_1=21$, $\Delta u_1=18$, $\Delta^2 u_1=-3$. Therefore

$$a=21a_{\overline{15}|\,\overline{1}|}+18a_{\overline{15}|\,\overline{2}|}-3a_{\overline{15}|\,\overline{3}|}$$

This is an annuity which increases till the maximum payment is 84, and which then decreases till the fifteenth payment vanishes.

27. A lease is granted for ten years at an annual rent, with power to the tenant to renew ten times for a like period on payment of a fine of 1 for each year of the lease expired. What is the value of the fines? Interest 5 per cent.

Here the fines are respectively 10, 20, etc., 100, payable at the end of 10, 20, etc., years. A sum payable periodically at the end of 10, 20, etc., years may, in accordance with Chapter II. Art. 30, be looked upon as an annuity at rate of interest $\{(1+i)^{10}-1\}=j$ say. If the annuity at rate j be denoted by $a'_{\overline{10}|}$, we have the value of the fines $10a'_{\overline{10}|\,\overline{1}|}+10a'_{\overline{10}|\,\overline{2}|}$. Calculating the values of these annuities when $i={\cdot}05$, and consequently $j={\cdot}6288946$, we shall find the value of the fines to be 39·66.

28. The following example is an instructive one:—

An annuity-certain deferred twenty years, and after that to run twenty years, is to be paid for by an annual premium, the first payment, π, to be paid down now, and afterwards, at the beginning of each year, a regularly diminishing amount, the last premium being paid at the beginning of the twentieth year, after which the premium becomes extinct. Find π, having given $v^{20}=\cdot456387$ at 4 per cent. interest.

Here the benefit to be received is an annuity for twenty years deferred twenty years. Its present value is therefore $v^{20}a_{\overline{20}|}$. The consideration for this benefit is an annuity for twenty years, commencing at π, and decreasing $\frac{\pi}{20}$ each year. If this annuity were payable at the end of each year, its value would be $\pi(a_{\overline{20}|\,\overline{1}|}-\frac{1}{20}a_{\overline{20}|\,\overline{2}|})$ but, as it is payable at the beginning of each year, its value is $\pi(1+i)\left(a_{\overline{20}|\,\overline{1}|}-\frac{1}{20}a_{\overline{20}|\,\overline{2}|}\right)$. The benefit and the consideration for the benefit must be equal in present value: whence

$$\pi(1+i)\left(a_{\overline{20}|\,\overline{1}|}-\frac{1}{20}a_{\overline{20}|\,\overline{2}|}\right)=v^{20}a_{\overline{20}|\,\overline{1}|}$$

$$\text{and } \pi=\frac{v^{20}a_{\overline{20}|\,\overline{1}|}}{(1+i)\left(a_{\overline{20}|\,\overline{1}|}-\frac{1}{20}a_{\overline{20}|\,\overline{2}|}\right)}$$

$$\begin{array}{rl}
v^{20}= & \cdot456387\\
1-v^{20}= & \cdot543613\div\cdot04\\
a_{\overline{20}|\,\overline{1}|}= & 13\cdot5903\\
20v^{20}= & 9\cdot1277\\
 & 4\cdot4626\div\cdot04\\
a_{\overline{20}|\,\overline{2}|}= & 111\cdot565\\
a_{\overline{20}|\,\overline{1}|}= & 13\cdot5903\\
\frac{1}{20}a_{\overline{20}|\,\overline{2}|}= & 5\cdot5783\\
 & 8\cdot0120\times1.04\\
 & 3205\\
(1+i)\left(a_{\overline{20}|\,\overline{1}|}-\frac{1}{20}a_{\overline{20}|\,\overline{2}|}\right)= & 8\cdot3325
\end{array}
\qquad
\begin{array}{r}
13\cdot5903\\
783654\\
\hline
54361\\
6795\\
815\\
41\\
10\\
1\\
\hline
v^{20}a_{\overline{20}|\,\overline{1}|}=6\cdot2023\\
8\cdot3325)\,6\cdot2023\,(\cdot74435\\
3695\\
362\\
29\\
4
\end{array}$$

$$\pi=\cdot74435$$

29. In former days, when De Moivre's hypothesis as to the law of life was of greater importance in the science of life contingencies than it now is, attention was much directed to the annuities-certain the payments of which are the powers of the natural numbers. Mr. Baily devoted two chapters of his work to this subject.

By means of the formulas of this chapter, the amounts and present values of such annuities can be readily found. Thus, let it be required to find the value of an annuity, the payments of which are 1^4, 2^4, 3^4, etc.

	Δ	Δ^2	Δ^3	Δ^4	Δ^5
Here $1^4=$ 1	15	50	60	24	0
$2^4=$ 16	65	110	84	24	
$3^4=$ 81	175	194	108		
$4^4=$ 256	369	302			
$5^4=$ 625	671				
$6^4=$1296					

Therefore the value of the annuity is

$$a_{\overline{n}|\,\overline{1}|}+15a_{\overline{n}|\,\overline{2}|}+50a_{\overline{n}|\,\overline{3}|}+60a_{\overline{n}|\,\overline{4}|}+24a_{\overline{n}|\,\overline{5}|}$$

To find the amount of an annuity for n years, the successive payments of which are n^3, $(n-1)^3$, $(n-2)^3$, etc.

Here $u_1=n^3$

$\Delta\, u_1=-(3n^2-3n+1)$

$\Delta^2 u_1=(6n-6)$

$\Delta^3 u_1=-6$

and $s=n^3 s_{\overline{n}|\,\overline{1}|}-(3n^2-3n+1)\,s_{\overline{n}|\,\overline{2}|}+(6n-6)s_{\overline{n}|\,\overline{3}|}-6s_{\overline{n}|\,\overline{4}|}$

30. Formulas (13), (14), and (15) apply only to annuities where the differences of the payments vanish after a finite number of orders, and it is not sufficient if the differences merely rapidly diminish but do not vanish. The $(r-1)^{th}$ difference is multiplied into an annuity of the rth order, and we have seen that the amounts and values of annuities of the higher orders are very large. Therefore, although the $(r-1)^{th}$ difference may be very small, its coefficient will be very large if r be large, and the product will be a quantity which cannot safely be neglected.

31. Let us take the following case :—

		Δ	Δ^2	Δ^3	Δ^4
1st payment.	1·000000	·010000	·000100	·000001	·000000
2d "	1·010000	·010100	·000101		
3d "	1·020100	·010201	etc.		
4th "	1·030301	etc.			
etc.	etc.				

These terms are in geometrical progression, the ratio being 1·01, and the differences diminish with considerable rapidity, the fourth having no significant figure in the sixth decimal place. Let it be required to find the value of the perpetuity at 5 per cent. by means of formula (15).

$$
\begin{array}{rl}
\Delta^3 u_1 = & \cdot000001 \div \cdot05 \\
 & \underline{\cdot000020} \\
\Delta^2 u_1 = & \underline{\cdot000100} \\
 & \underline{\cdot000120} \div \cdot05 \\
 & \cdot002400 \\
\Delta u_1 = & \underline{\cdot010000} \\
 & \underline{\cdot012400} \div \cdot05 \\
 & \cdot248000 \\
u_1 = & \underline{1\cdot000000} \\
 & \underline{1\cdot248000} \div \cdot05 \\
 & \underline{\underline{24\cdot960000}}
\end{array}
$$

The truo value of the perpetuity is 25·0, so that although the first difference neglected is only ·0000001, there is a considerable error in our result. This is accounted for by the fact that the coefficient of the first neglected difference is $\frac{1}{i^5}$ or 3,200,000. Of course theoretically any required degree of accuracy could be obtained by computing a sufficient number of terms to a sufficient number of decimal places, provided that the differences of the series of payments diminish faster than the increase in the coefficients; but more convenient formulas for such cases we now proceed to find.

32. From the example in the last article, we infer that formulas (13), (14), and (15), are not applicable to annuities which increase or decrease in geometrical progression; but it is easy to give a general demonstration of the fact.

Let there be a geometrical series, the first term of which is K and common ratio R. The following scheme shows the series and its differences:—

Term	Δ	Δ^2	Δ^3	Δ^4
K	$K(R-1)$	$K(R-1)^2$	$K(R-1)^3$	$K(R-1)^4$
KR	$KR(R-1)$	$KR(R-1)^2$	$KR(R-1)^3$	etc.
KR^2	$KR^2(R-1)$	$KR^2(R-1)^2$	etc.	
KR^3	$KR^3(R-1)$	etc.		
KR^4	etc.			
etc.				

It therefore appears that if the common ratio of the original series be R, the successive orders of differences form another geometrical series with the common ratio $(R-1)$ and the differences can never vanish. Therefore, by Art. 30, formulas (13) to (15) are inapplicable.

33. To find the value of an annuity, the payments of which are in geometrical progression with common ratio R.

If a be the value of the annuity, we have

$$a=(1+i)^{-1}+R(1+i)^{-2}+R^2(1+i)^{-3}+\text{ etc. }+R^{n-1}(1+i)^{-n}$$

$$=\frac{1}{R}\{R(1+i)^{-1}+R^2(1+i)^{-2}+R^3(1+i)^{-3}+\text{etc }+R^n(1+i)^{-n}\}$$

$$=\frac{1}{R}\cdot\frac{R}{1+i}\,\frac{1-\left\{\dfrac{R}{(1+i)}\right\}^n}{1-\dfrac{R}{(1+i)}}$$

$$=\frac{1-\left\{\dfrac{R}{1+i}\right\}^n}{(1+i)-R}\quad . \qquad (16)$$

Let $\dfrac{R}{(1+i)}=\dfrac{1}{(1+j)}$, so that $j=\dfrac{(1+i)-R}{R}$,

Then

$$a=\frac{1}{R}\cdot\frac{1-(1+j)^{-n}}{j}\quad . \qquad (17)$$

If the annuity be a perpetuity, we can find its present value only when $\dfrac{R}{1+i}<1$, that is, when $R<(1+i)$. In other cases the present value will be infinite. When $R<(1+i)$ and the annuity is a perpetuity, then

$$a=\frac{1}{(1+i)-R}\quad . \qquad (18)$$

In the example in Art. 31, $R=1{\cdot}01$, and the value of the perpetuity at 5 per cent. is therefore, by formula (18), $\dfrac{1}{{\cdot}04}$ or 25. Thus the value of the increasing perpetuity is equal to the value of a fixed perpetuity of 1 calculated upon the assumption that the rate of interest is diminished by the rate of increase of the payments.

34. From formula (17) we see that by changing the rate of interest, we can substitute for an annuity, the payments of which are in geometrical progression, another annuity, the payments of which are uniform. The changed rate of interest may, however, present anomalous features. If R be greater than unity, then j, the substituted rate, will be less than i; and if R be greater than

$(1+i)$, then j will be negative. If R be less than unity, then j will be greater than i; and there is no limit to the magnitude which j may assume when R is diminished.

35. It is to Mr. William M. Makeham that we owe those formulas in this chapter, by which, with the aid of the figurate numbers, we can deal so effectively with variable annuities. Previous writers had investigated particular cases, but Mr. Makeham has furnished the general theory. He published it in a remarkable paper in *J. I. A.* vol. xiv. p. 189.

CHAPTER IV.

On Loans Repayable by Instalments.

1. In Chapter II. we gave full consideration to loans which are repayable within a limited term by equal periodic instalments, including principal and interest. It is the object of the present chapter to extend our investigation to cases in which the capital is repayable in any other manner whatever. The formulas of Chapter III. will be of great value to us in our inquiry.

The analysis divides itself naturally into two branches which are of equal importance; namely, first, knowing the conditions of the loan, to find that value for it which will secure to an investor a given rate of interest; and secondly, to ascertain the rate of interest which the loan yields at a given market price.

2. When a corporation or a foreign government contracts a loan, it generally undertakes obligations of a twofold nature. It agrees to pay the lender interest at a fixed rate as long as his advance remains outstanding, and it promises to repay at stated periods the capital itself. Frequently the borrower holds out to investors inducements beyond the stipulated rate of interest, by issuing the loan at a discount and repaying it at par, or by issuing it at par and giving a bonus on repayment. Sometimes, on the other hand, where the credit of the borrower is good, he may find it to his advantage to raise the loan at a higher rate of interest than the public demands from him, and therefore to issue his bonds at a premium. These complications, however, will not render our investigations much more intricate.

3. Let C = the capital repayable by the borrower.

j = the nominal rate of interest thereon paid by the borrower.

i = the actual rate of interest realized by the lender, whom we may also call the investor, or purchaser.

K = the present value of the capital at rate i.

A = the purchase-money, or the value of the loan.

To illustrate the symbols, let us suppose a loan to be contracted for £10,000,000 at 3 per cent., by ten thousand bonds of £1000 each, the bonds to be paid off at maturity with a bonus of 25 per cent. Here the capital repayable by the borrower is really £12,500,000, which sum we therefore represent by C, and the nominal rate of interest is not ·03, as would appear from the stated conditions, but $\frac{3}{125}$, or ·024, which we represent by j. The loan being nominally issued at par, we have $A=10,000,000$.

The points to be noted in connection with our symbols are, that C represents the capital repayable by the borrower, including any bonus he may contract to pay along with it, and that j represents the ratio between the annual interest contracted for and the capital repayable as defined above. We must therefore be careful to distinguish, as in the foregoing example, between the nominal rate of interest as stated in the conditions of the loan, and the nominal rate j which concerns us in our investigations. These rates will in many cases be identical, but they are not necessarily so.

4. To find the value of a loan, repayable by instalments at stated periods of time, with interest in the meantime at rate j, so as to yield the purchaser a given rate of interest, i.

The value of the loan consists of two parts, the value of the capital and the value of the interest. The value of the capital at rate i being K, the value of the interest is evidently $(A-K)$. Had the borrower contracted to pay interest at rate i, then the loan would have stood at par, and A would have been equal to C; and therefore the value at rate i of the interest in this particular case would have been $(C-K)$. The annual interest on the loan would, on the same supposition, have been iC, and the value of each annual unit of interest payable by the borrower is therefore $\frac{C-K}{iC}$. But the borrower has actually contracted to pay interest at rate j instead of i, or jC annually, and the value of this interest is $\frac{jC}{iC}(C-K)$, or $\frac{j}{i}(C-K)$. Adding to this the value of the capital, we have the value of the whole loan,

$$A=K+\frac{j}{i}(C-K) \quad . \quad (1)$$

5. It will perhaps enable us better to understand the demonstration of the last article if we confine our attention for a moment to a single unit of the loan. Suppose that unit to be represented by C_1, and to be payable at the end of n_1 years, and let its value be

K_1, so that $K_1=(1+i)^{-n_1}$. The interest payable on the unit is an annuity of j per annum for n_1 years; but the value of 1 per annum for n_1 years is $\frac{1-(1+i)^{-n_1}}{i}$, or $\frac{C_1-K_1}{i}$, and therefore the value of j per annum is $\frac{j}{i}(C_1-K_1)$. Adding to this the value of the unit of the capital itself, we have $\left\{K_1+\frac{j}{i}(C_1-K_1)\right\}$, the total value of that particular loan-unit under consideration.

If now we take into account the other units C_2, C_3, etc., due at the end of n_2, n_3, etc., years, we have corresponding formulas $\left\{K_2+\frac{j}{i}(C_2-K_2)\right\}$, $\left\{K_3+\frac{j}{i}(C_3-K_3)\right\}$, etc., and, adding together the values of all these portions of the loan, we have for the value of the entire loan

$$(K_1+K_2+K_3+\text{etc.})+\frac{j}{i}\left\{(C_1+C_2+C_3+\text{etc.})-(K_1+K_2+K_3+\text{etc.})\right\}$$

or $\left\{K+\frac{j}{i}(C-K)\right\}$ as before.

6. In Art. 4 we have treated the capital of the loan as a whole. For purposes of calculation it will often be convenient to represent the entire capital by unity, so that we write $C=1$. Under these circumstances formula (1) takes the elegant form

$$A=K+\frac{j}{i}(1-K)$$
$$=1-(1-K)\left(1-\frac{j}{i}\right) \quad . \quad . \quad (2)$$

7. In formulas (1) and (2) we do not limit the repayments of capital to any particular conditions. We simply find the present value, K, of the capital, under whatsoever arrangements it may be repayable, and insert it in the formulas. The formulas will take various shapes according to the various methods by which loans may be liquidated. Thus, if the loan of unity is to be repaid in one sum at the end of n years, we have

$$A=1-(1-v^n)\left(1-\frac{j}{i}\right)$$
$$=1-a_{\overline{n}|}(i-j) \quad . \quad . \quad . \quad . \quad . \quad (3)$$

where $a_{\overline{n}|}$ is taken at rate i.

Again, if the capital of 1 be repayable by n equal annual instalments of $\frac{1}{n}$ each, then

$$A=1-\left(1-\frac{a_{\overline{n}|}}{n}\right)\left(1-\frac{j}{i}\right) \quad . \quad . \quad (4)$$

Also, if the capital of C be repayable by annual instalments, u_1 at the end of the first year, u_2 at the end of the second year, and so on, then in formula (1), $K=vu_1+v^2u_2+v^3u_3+$ etc., and if the payments u_1, u_2, etc., form a series, the differences of which vanish after a finite number of orders, then by Chapter III., formula (14),

$$K=a_{\overline{n}|\,\overline{1}|}\,u_1+a_{\overline{n}|\,\overline{2}|}\,\Delta u_1+a_{\overline{n}|\,\overline{3}|}\,\Delta^2 u_1+\text{ etc.} \qquad (5)$$

8. The following examples will elucidate the subject:—

(*a.*) A bond for £1000 is to be sold. It bears interest at 3 per cent., and will be repaid at par in twenty years. An intending purchaser desires to make 5 per cent. on his investment. What price can he afford to give for the bond? Here formula (3) is applicable, writing $a_{\overline{n}|}=a_{\overline{20}|}$ at 5 per cent., and $i=\cdot05$ and $j=\cdot03$.

$$A=1000\{1-a_{\overline{20}|}\,(\cdot05-\cdot03)\}$$
$$=1000\{1-12\cdot462\times\cdot02\}$$
$$=750\cdot76.$$

(*β.*) A loan of £1000 is to be repaid as follows, with a bonus of 25 per cent.:—

One twenty-seventh at the end of four years,
" " five years,
and so on. Finally,
one twenty-seventh at the end of thirty years,

interest being payable in the meantime at the rate of 6 per cent. What price must a purchaser give so as to realize 5 per cent. on his outlay?

Here in formula (1) we must write $C=1250$ and $j=\frac{6}{125}$, or $\cdot048$; also, $K=\frac{v^3a_{\overline{27}|}}{27}\times1250$, at 5 per cent.

$$\begin{aligned}
\log v^3 &= \bar{1}\cdot93643\\
\log a_{\overline{27}|} &= 1\cdot16563\\
\log 1250 &= 3\cdot09691\\
&\overline{\;4\cdot19897\;}\\
\log 27 &= 1\cdot43136\\
\log K &= \underline{2\cdot76761}\\
K &= 585\cdot61\\
\tfrac{j}{i}(C-K) &= \underline{637\cdot82}\\
A &= \underline{1223\cdot43}
\end{aligned}
\qquad
\begin{aligned}
C-K &= 664\cdot39\\
&840\cdot\\
&26576\\
&5315\\
\cdot05\,)\,&\overline{31\cdot891}\\
\tfrac{j}{i}(C-K) &= 637\cdot82
\end{aligned}$$

γ. A loan of £10,000 is repayable as follows:—

£94 at the end of the first year,
102 " second year,
110 " third year,

and so on till the whole loan is liquidated, interest at the rate of 3 per cent. being allowed in the meantime on the outstanding capital. Required the value of the loan at 5 per cent. We must first find the term of the loan. By the well-known formula for summation in the calculus of Finite Differences, if there be a series of n terms, the first of which is u_1, the sum of the series is $nu_1+\frac{n(n-1)}{\underline{|2}}\Delta u_1+$ etc. In the present case the sum of the series is 10,000, u_1 is 94, and Δu_1 is 8. Making use of these values, we have a quadratic equation in n, namely, $4n^2+90n-10,000=0$; whence $n=40$. The capital repayments are therefore of the nature of an increasing annuity for forty years—first payment 94, second payment 102, third payment 110, etc., and by formula (5) we have $K=94a_{\overline{40}|\,\overline{1}|}+8a_{\overline{40}|\,\overline{2}|}$. In Chapter III., Art. 18, we have already calculated the value of $a_{\overline{40}|\,\overline{2}|}$, and employing those figures, we have $K=3449{\cdot}316$. Making use of this value in formula (1),

$$A=3449{\cdot}316+\frac{3}{5}\times 6550{\cdot}684$$
$$=7379{\cdot}726,$$

or £73, 15s. 11¼d. per cent.

9. In the foregoing investigation we have divided the loan into two parts, the capital and the interest, and we have expressed the value of the interest in terms of the value of the capital. We might have pursued a different, and apparently a more direct course, and valued the capital and interest separately, but the effect of the method of solution which we have followed, and which is due to Mr. W. M. Makeham, is generally to reduce the series to be valued by one order of differences. Thus, suppose we take the case of the capital of unity repayable by n equal annual instalments of $\frac{1}{n}$ each, —formula (4)—and value the capital and interest independently. The value of the capital is $\frac{a_{\overline{n}|}}{n}$. The interest payable at the end of the first year is j, at the end of the second year $\left(1-\frac{1}{n}\right)j$, at the end of the third year $\left(1-\frac{2}{n}\right)j$, and so on. The value of the interest is therefore $ja_{\overline{n}|\,\overline{1}|}-\frac{j}{n}a_{\overline{n}|\,\overline{2}|}$, and the value of the whole loan $A=a_{\overline{n}|}\left(\frac{1}{n}+j\right)-\frac{ja_{\overline{n}|\,\overline{2}|}}{n}$. This formula, although it involves an annuity of the second order, is really identical with formula (4).

Thus,

$$
\begin{aligned}
&a_{\overline{n}|}\left(\frac{1}{n}+j\right)-\frac{ja_{\overline{n}|\,2|}}{n}\\
&=a_{\overline{n}|}\left(\frac{1}{n}+j\right)-\left(\frac{a_{\overline{n}|}}{i}-\frac{n(1+i)^{-n}}{i}\right)\frac{j}{n}\\
&=a_{\overline{n}|}\left(\frac{1}{n}+j-\frac{j}{ni}\right)+(1+i)^{-n}\times\frac{j}{i}\\
&=a_{\overline{n}|}\left(\frac{1}{n}+j-\frac{j}{ni}\right)-ja_{\overline{n}|}+\frac{j}{i}\\
&=\frac{a_{\overline{n}|}}{n}\left(1-\frac{j}{i}\right)+\frac{j}{i}\\
&=1-\left(1-\frac{a_{\overline{n}|}}{n}\right)\left(1-\frac{j}{i}\right),\text{ as in formula (4).}
\end{aligned}
$$

10. We pass now to the converse problem:—Having given the value of the loan, to determine the rate of interest.

11. We have by formula (1), $A=K+\frac{j}{i}(C-K)$, whence, by simple algebraical transformation,

$$i=j\frac{C-K}{A-K} \qquad \text{(6)}$$

In this equation, seeing that i is the unknown quantity, and that K is calculated at rate i, we cannot assign the true value to K, but if we find an approximate value for K, and insert it in the formula, we shall obtain an approximate value for i; and the nearer the assumed value is to the true value of K, the nearer will the resulting approximate value be to the true value of i.

12. As an illustration of the formula, let us take the following example. What rate of interest does a Government loan issued at 73 per cent. yield when it is redeemable at par by uniform annual drawings of 2 per cent.?

Here the 3 per cent. paid by the Government yields a little over 4 per cent. on the issue price of 73, and, in addition, the lender will on repayment get a bonus of 27 on each 73 invested, and this is evidently equal to fully 1 per cent. per annum additional interest. We may therefore assume, to begin with, a rate of 5 per cent. The formula becomes $i = \cdot 03 \times \dfrac{100-2a_{\overline{50}|}}{73-2a_{\overline{50}|}}$, where $a_{\overline{50}|}$ is taken as a trial at 5 per cent. The result is, $i = \cdot 03 \times \dfrac{61\cdot 488}{34\cdot 488} = \cdot 05349$.

The assumed rate turns out to be too low, and, to get a closer approximation, we might insert in the formula the value just found for i, and work it out again. We can, however, proceed in another way, which will often, by the help of interest tables, be more easily applied.

Assuming a higher rate, say $5\frac{1}{2}$ per cent., we find i to be ·05070, and from the two approximate rates for i we can find a third more near than either. Thus

5	gives	5·349	per cent.
5·5	gives	5·070	per cent.
Diff. ·5	gives	Diff. −·279	

whence $5+x=5\cdot349-\frac{\cdot279}{\cdot5}x$, nearly,

whence $x=\cdot224$, and the rate which we seek is 5·224 per cent. very nearly.

13. The rationale of the method of approximation above illustrated is easily seen. If the trial rate, which we may denote by I_1, were the true one, then that rate itself would be the result of making use of it in the formula, but, seeing that I_1 is only approximately true, our result, which may be written J_1, is also only approximate. If now we assume another trial rate, I_2, differing from I_1 by Δ, say, we get another approximate result, J_2, differing from J_1 by δ, say. Since a change of Δ in the trial rate produces a change of δ in the resulting rate, therefore a change of x in the trial rate will produce a change of $\frac{\delta}{\Delta}x$ in the resulting rate. We seek a rate which, when used in the formula, will reproduce itself, and, if that rate be I_1+x, we must therefore have $I_1+x=J_1+\frac{\delta}{\Delta}x$,

whence $x=\frac{\Delta\times(J_1-I_1)}{\Delta-\delta}$ and

$$i=I_1+\frac{\Delta\times(J_1-I_1)}{\Delta-\delta} \qquad (7)$$

This method of approximation may often be advantageously resorted to with other formulas than that to which we have just now applied it.

14. It is very easy to show that formula (6) of the present chapter is an extension and generalization of Baily's, No. 35 of Chapter II., which applies only where the loan is repayable by equal annual instalments, including principal and interest. In the case of such a loan we have, by Chapter II., Art. 34, the capital included in the mth payment equal to v^{n-m+1}. The present value of the capital in the mth payment is $v^m v^{n-m+1}$, or v^{n+1}, or $(1+i)^{-(n+1)}$; and there being n such payments, the present value of the whole of

See page 243 Art 9.

the capital is $n(1+i)^{-(n+1)}$. Now, if an appropriate rate I be obtained by inspection of the tables, we have in formula (6) $C=\frac{1-(1+I)^{-n}}{I}$, and $K=n(1+I)^{-(n+1)}$, while A is the value of the loan.

Therefore
$$i=I\times\frac{\frac{1-(1+I)^{-n}}{I}-n(1+I)^{-(n+1)}}{A-n(1+I)^{-(n+1)}}$$
$$=I\left\{1+\frac{\frac{1-(1+I)^{-n}}{I}-A}{A-n(1+I)^{-(n+1)}}\right\}$$
whence
$$i-I=I\times\frac{\frac{1-(1+I)^{-n}}{I}-A}{A-n(1+I)^{-(n+1)}}. \quad (8)$$

In Chapter II., formula (35), we have denoted $(i-I)$ by ρ, I by j, $\frac{1-(1+I)^{-n}}{I}$ by a', A by a, and $n(1+I)^{-(n+1)}$ by nv^{n+1}. Making these substitutions in the above formula (8), we at once have formula (35) of Chapter II.

15. Just as Barrett's formula—No. (38) of Chapter II.—is an improvement on Baily's, so we may obtain an improvement on No. (6) of this chapter. We have seen, Art. 4, that $(C-K)$ is the value of the interest when it is payable at rate i. Therefore $\frac{1}{i}(C-K)$ is the value of the interest for each unit of the rate, and $\frac{i}{(C-K)}$ is the annual rate of interest for which a payment of 1 down will provide. Hence $i\frac{C-A}{C-K}$ is the extra annual rate of interest for which the discount $(C-A)$ will provide, and this extra rate, which we may denote by h, must be added to j, the rate actually payable, in order to get i, the rate realized by the lender. Seeing that i is unknown, we may take a near value, and denoting that by I, we have approximately

$$h=I\frac{C-A}{C-K} \quad (9)$$

16. We have shown that formula (6) is a generalization of Baily's No. (35) of Chapter II., and now we can similarly show that formula (9) is a generalization of Barrett's No. (38) of Chapter II. As in Art. 14, we have C equivalent to a', A equivalent to a, and K equivalent to nv^{n+1}, while we have now denoted by h that which we formerly wrote ρ, and by I that which

we formerly wrote j. Formula (9) therefore becomes $\rho=j\frac{a'-a}{a'-nv^{n+1}}$ which is identical with Barrett's formula.

17. Following precedent, we shall close this chapter with a few examples.

(*a*.) A loan of £2,000,000 is issued, repayable as follows:—

50,000 at the end of 5 years.
60,000 ,, ,, 6 ,,
70,000 ,, ,, 7 ,,

and so on till all is repaid, interest on the outstanding amount being allowed at the rate of 5 per cent. The issue price of the loan is 93 per cent. What does the loan cost the borrower?

Here, by a process similar to that followed in example γ of Art. 8, we find that the loan will all be paid off in 16 years after the repayments commence. The repayments are in the form of an annuity, and as the first payment of that annuity will take place at the end of five years, the annuity is deferred four years. We therefore have for the value of the capital, after dividing by 10,000 for the sake of brevity, $K=v^4(5a_{\overline{16}|}+a_{\overline{16}|\,\overline{2}|})$, while $C=200$, and $A=186$. If we use formula (9) and take for a trial rate $5\frac{1}{2}$ per cent., we have

$$a_{\overline{16}|}=10{\cdot}46216$$
$$16v^{16}=\underline{\ 6{\cdot}79330}$$
$$\cdot055)\ 3{\cdot}66886\ (66{\cdot}706=a_{\overline{16}|\,\overline{2}|}$$

368
388
360

$$5a_{\overline{16}|}=\ 52{\cdot}311$$
$$a_{\overline{16}|\,\overline{2}|}=\underline{\ 66{\cdot}706}$$
$$119{\cdot}017$$
v^4 reversed 712708

952136
8331
238
12
8

$$K=96{\cdot}0725$$
$$h=\cdot007401$$
$$j=\underline{\cdot05}$$
$$i=\underline{\cdot057401}=5{\cdot}7401 \text{ per cent.}$$

$$h=\cdot055\times\frac{200-186}{200-96{\cdot}0725}$$
$$=\cdot055\times\frac{14}{103{\cdot}9275}$$
$$\log\cdot055=\bar{2}{\cdot}7404$$
$$,,\ 14=\underline{1{\cdot}1461}$$
$$\bar{1}{\cdot}8865$$
$$\log 103{\cdot}9275=\underline{2{\cdot}0166}$$
$$\log h=\bar{3}{\cdot}8699$$

The trial rate of $5\frac{1}{2}$ per cent. is evidently too low. If we try again at 6 per cent. we have $a_{\overline{16}|\,\overline{2}|}=63{\cdot}459$ and $K=90{\cdot}2899$, and the rate brought out by the calculation will be 5·7656. Applying now formula (7) we have $I_1=5{\cdot}5$ per cent., $J_1=5{\cdot}7401$ per cent., $\Delta={\cdot}5$, and $\delta={\cdot}0255$, whence the final rate $i=5{\cdot}753$ per cent.

(*β.*) Colonial 5 per cent. Government bonds repayable at par in 19 years are quoted in the market at $107\frac{3}{4}$ per cent., after making allowance for the interest accrued since the last payment of dividend. What rate of interest do they yield? and, to yield the same return, what should be the price of 4 per cent. bonds repayable at par in 25 years?

For the first part of the question, using again formula (9) and trying 4 per cent., we have (K being equal to 47·464), $i=4{\cdot}410$ per cent. Trying again $4\frac{1}{2}$ per cent. we have (K being equal to 43·330), $i=4{\cdot}385$ per cent. Interpolating by means of formula (7) we finally have for the rate yielded by the bonds 4·390 per cent.

For the second part of the question we may conveniently employ formula (2) where $j={\cdot}04$ and $i={\cdot}0439$, and where K is v^{25} taken at rate i, or ·34161. The formula becomes $A=1-{\cdot}65839\times\left(1-\frac{4}{4{\cdot}39}\right)$ or 94·151 per cent.

18. In practically applying the formulas of this chapter and of those that precede, circumstances may render necessary various modifications; and it may also frequently happen that there is no formula given which will directly meet the case in hand. The principles will however remain constant, and the actuary who has fully mastered the principles will find no difficulty in adapting the formulas to special conditions. In the affairs of life mathematical rules cannot be made rigidly to apply, and the actuary, having made himself thoroughly acquainted with the mathematical rules, must never fail to exercise a sound judgment when he comes to make use of them.

$$A'_{xy} = \tfrac{1}{2}\left\{A_{xy} + \frac{a_{x-1:y}}{p_{x-1}} - \frac{a_{x:y-1}}{p_{y-1}}\right\}.$$

$$= \tfrac{1}{2}\left\{A_{xy} + \frac{v p_{x-1:y}\, a_{x:y+1}}{p_{x-1}} - \frac{v p_{x:y-1}\, a_{x+1:y}}{p_{y-1}}\right\}$$

$$= \tfrac{1}{2}\left\{A_{xy} + v p_y\, a_{x:y+1} - v p_x\, a_{x+1:y}\right\}$$

$$= \tfrac{1}{2}\left\{1 - d\, A_{xy} + v p_y\, a_{x:y+1} - v p_x\, a_{x+1:y}\right\}.$$

$$\frac{-v p_{x-1}}{p_{x-1}}$$

$$A_{xy\,\overline{n|}} = A_{x\,\overline{n|}} + A_{y\,\overline{n|}} - A_{xy\,\overline{n|}}.$$

$$A_{\overline{xy}\,\overline{n|}} = A_{x\,\overline{n|}} + A_{y\,\overline{n|}} - A_{\overline{xy}\cdot\overline{n|}}$$

$A_{\overline{xy}\,\overline{n|}}$ is a small value & may be neglected.

$$P'_{\overline{xy}\,\overline{n|}} = P_{y\,\overline{n|}} + \pi_{x\,\overline{n|}}(1 + \phi$$

THEORY OF FINANCE.

CHAPTER V.

On Interest Tables.

1. In the preceding chapters we have given formulas by means of which all the values to be found in interest tables could be independently calculated as required, but it is evident that such a process would be tedious, and that the convenience is great of having those values that will be commonly wanted ready prepared and presented in tabular form. The solution of many problems too—such, for instance, as finding the rate of interest involved in a term annuity—is rendered very much more easy by a reference to tables, and to the skilful actuary they are at all times invaluable auxiliaries in his work. It will be frequently noticed that where a novice goes through an intricate calculation in answering a question, the adept produces the same result seemingly without thought or effort; and on inquiry it will usually be discovered that tables are the tools he uses to shorten his labour and save his time.

2. An intimate acquaintance with the nature of the tables he may find in his hands is essential to the actuary, for without that knowledge he cannot turn them to the best account. It is therefore very desirable for him to practise the construction of tables for himself, although those he requires may already be in print, as by actual experience in their manufacture he will much more readily obtain a clear knowledge of the properties of his tools, than by only theoretical study. The principal object of the present chapter is therefore to explain the best methods of constructing and verifying interest tables, and it will for the most part be left to the reader to gain for himself practical skill in using them, by studying the first four chapters of the work with the tables in his hands. Sufficient tables are given at the end to enable him to do so without going beyond the pages of the book itself.

3. In interest tables the functions most commonly tabulated for each rate of interest are the following:—

(i.) $(1+i)^n$, the amount of 1 in n intervals.

(ii.) v^n, the present value of 1 due at the end of n intervals.

(iii.) $s_{\overline{n}|}$, the amount of an annuity of 1 for n intervals.

(iv.) $a_{\overline{n}|}$, the present value of an annuity of 1 for n intervals.

(v.) $\frac{1}{s_{\overline{n}|}}=s^{-1}_{\overline{n}|}$, the sinking fund which will redeem a debt of 1 in n intervals; or, in other words, the annuity which will accumulate to 1 in n intervals.

(vi.) $\frac{1}{a_{\overline{n}|}}=a^{-1}_{\overline{n}|}$, the annuity for n intervals which 1 will purchase.

The values under each heading are arranged in columnar form, commencing with the value for one interval, and finishing generally with that for one hundred intervals.

4. The following is an example of the form which the interest table sometimes assumes:—

INTEREST 5 PER CENT.

n	(i.) $(1+i)^n$	(ii.) v^n	(iii.) $s_{\overline{n}\|}$	(iv.) $a_{\overline{n}\|}$	(v.) $s^{-1}_{\overline{n}\|}$	(vi.) $a^{-1}_{\overline{n}\|}$
1	1·050000	·952381	1·0000	·9524	1·000000	1·050000
2	1·102500	·907029	2·0500	1·8594	·487805	·537805
3	1·157625	·863838	3·1525	2·7232	·317209	·367209
etc.	etc.	etc.	etc.	etc.	etc.	etc.

5. All tables are not arranged in the same manner. Frequently the functions, all at one rate of interest, are placed in parallel columns as in the specimen in Art. 4, so that there is a distinct table for each rate of interest. Sometimes each function is kept by itself, the rates of interest being side by side. Such are the tables which are given at the end of the book.

6. Corbaux in his work, published in 1825, *Doctrine of Compound Interest*, supplies very complete tables. He gives all the columns (i.) to (vi.) mentioned in Art. 4, for each rate of interest rising by $\frac{1}{4}$ per cent. from 3 per cent. to 6 per cent.; and not only does he do so, but for each rate of interest he also gives the values of all the

functions when interest is convertible either half-yearly or quarterly.

Some tables, instead of supplying the values for interest convertible half-yearly or quarterly, very minutely subdivide the rate of interest, and also begin with very small rates. In this way they answer the same purpose as Corbaux's tables. Thus an annuity of 100 per annum payable quarterly for twenty years, at $4\frac{1}{2}$ per cent., interest convertible quarterly, is equivalent to an annuity of 25 per annum for 80 years at $1\frac{1}{8}$ per cent., interest convertible yearly. We therefore see that in speaking of interest tables, it it more appropriate to name *intervals*, as we have done, rather than *years*.

The tables of Colonel Oakes, published in 1877, are of the last-named description. They furnish columns (i.) to (iv.) at each rate of interest, rising $\frac{1}{8}$ per cent. from $\frac{3}{4}$ per cent. to 10 per cent. Such also are the tables of the late Mr. P. Hardy, F.R.S., published in 1839. They contain columns (i.) to (iv.) for rates beginning at $\frac{1}{4}$ per cent., and rising by $\frac{1}{4}$ per cent. to 5 per cent., and also for 6, 7, and 8 per cent. Rance's, published in 1852, are of the same kind, and give cols. (i.) to (iv.) for rates beginning at $\frac{1}{4}$ per cent., and proceeding by steps of $\frac{1}{4}$ per cent. to 10 per cent.

7. It may be noted here that although some published tables give all the six columns (i.) to (vi.), it is not really necessary that columns (v.) and (vi.) should both appear. We saw, Chapter II., formula (29), that $a_{\overline{n}|}^{-1}-i=s_{\overline{n}|}^{-1}$. If therefore $a_{\overline{n}|}^{-1}$, the annuity which 1 will purchase, be given, we can at once, by deducting the rate of interest, find the sinking fund. We therefore, in the specimen tables at the end of the book, have not included column (v.) of Art. 4.

8. We have said that it is usual to tabulate the functions for all values of n from 1 to 100, but for ordinary purposes a table of such extent is not essential. It is not often in practice that the values and amounts of annuities are required for so long a period as 100 years. A table, too, may be effectually used for longer terms of years than it actually includes. Thus, suppose there is a table which goes up to 50 years only, and it is required to find the values of the functions for 60 years. We have $(1+i)^{60}=(1+i)^{50}\times(1+i)^{10}$, $v^{60}=v^{50}\times v^{10}$, $a_{\overline{60}|}=a_{\overline{50}|}+v^{50}a_{\overline{10}|}$, and $s_{\overline{60}|}=(1+i)^{10}s_{\overline{50}|}+s_{\overline{10}|}$. It is, however, convenient to possess extensive tables, especially if they are of the description of those of Oakes, and not of Corbaux, and we must use them for intervals instead of years; because, for example, an annuity for 25 years at

5 per cent., payable quarterly, is equivalent to one at $1\frac{1}{4}$ per cent. for 100 intervals.

9. Insurance premiums are payable in advance, and thus are of the nature of annuities-due. It might therefore be thought useful for the purposes of life offices to tabulate the values and amounts of annuities-due instead of those of ordinary annuities: but the more usual form of tables, which is that we have adopted, provides for every object. Thus the amount of an annuity-due for n years is evidently $s_{\overline{n+1}|}-1$, which can be found by inspection from a table of the amounts of ordinary annuities. For instance, to find the amount of an annuity-due for 25 years at 4 per cent. interest, we enter table 3 with 26 years, and, deducting unity, the result is 43·31174. Again, to find the sinking fund payable at the beginning of each year to redeem a debt of 1 in n years, we can find the sinking fund payable at the end of each year and multiply it by v. For example, to find at 4 per cent. interest the sinking-fund payable at the beginning of each year which will redeem 1 in 25 years, we use table 5, and find first the annuity which 1 will purchase for 25 years. This is ·064012. Deducting the rate of interest, ·04, we have ·024012, the sinking-fund payable at the end of each year to redeem 1 in 25 years. Multiplying now by ·961538, the value of v taken from table 2, we have ·023088, the quantity required. This might be obtained equally easily in another way, namely, by finding the amount of an annuity-due for 25 years, and taking the reciprocal by means of Barlow's or Oakes's tables of reciprocals.[1]

10. To construct a table, the most obvious course is to calculate independently each value that is to be tabulated; but that course would very seldom be the best to pursue. It would usually be very laborious; and, besides, in order to insure accuracy, the whole work would have to be done in duplicate. It is generally much preferable to take some formula connecting the consecutive values of the function, and by means of it to compute them one from the other in succession. In this way each value is made to depend on all that go before it, with the consequence that if an error occur in one it is carried on to those that succeed, and we can therefore feel confidence that if any particular value be correct, all those that go before it are correct also. This method of constructing tables is

[1] By means of Orchard's Tables, the sinking fund payable at the beginning of each year to redeem a debt in n years, can be found by inspection. We have only to enter the table of annual premiums with $a_{\overline{n-1}|}$, and the result is the sinking fund required. It is beyond the scope of the present work to describe Orchard's Tables.

called the "continuous method," and when it is used, a periodical check, say at every tenth value, is all that is required.

11. To employ the continuous method three things are necessary. We must have a convenient *working formula* connecting the value of the function for n years with that for $(n+1)$ years: we must know the *initial value* on which all the others are to be built: and we must have a *verification formula* by which to apply our periodical checks. In interest tables these three requisites are simple, and easily obtained. In many other tables they do not present themselves so obviously, but the computer will find it to his advantage to seek them in order to construct his tables continuously, on account of the great facilities which the continuous method gives for insuring accuracy.

12. To construct a table of $(1+i)^n$. This table, for the majority of the rates of interest in use, can best be formed by direct multiplication. The values in the column are connected by the relation $(1+i)^{n+1}=(1+i)\,(1+i)^n$, and i being a small quantity not usually consisting of many digits, multiplication by $(1+i)$ is easy. We add to $(1+i)^n$ the result of its multiplication by i, and so obtain $(1+i)^{n+1}$. This is our working formula. The amount of 1 in one year is $(1+i)$, our initial value. To check our work we must calculate by means of logarithms, say every tenth value; or, if we construct the tenth value by logarithms, we can form the twentieth, thirtieth, etc., by raising the tenth to the second, third, etc., powers by ordinary contracted multiplication.

13. When i consists of but one significant figure—for instance, when $i=\cdot04$—the work of making the table is very easy. A type of the operation is given in the margin. The number of decimal places will go on increasing indefinitely unless the increase be checked. When we have obtained as many as we require, we must, as in the example, cease to allow them to extend, merely taking account of the proper carriage from the neglected figures to the figures which we retain. Thus, in the marginal example, the result of multiplying the last two figures of $(1+i)^3$ by $\cdot04$ is 256: we neglect the 56, and as the figures neglected are *greater than* 49, we increase the next figure by unity, carrying 3 instead of 2. In order to insure accuracy in the last decimal place of the tables, we must work to two

	$(1+i)^n$ 4 per cent.
1.	1·04
	416
2.	1·0816
	43264
3.	1·124864
	44995
4.	1·169859
	46794
5	1·216653
*	* *

places more than we mean finally to keep, and when our work is finished we must cut down the results to the required limit, taking care always, when the value of the rejected figures is greater than 49, to increase by unity the last place retained. Thus, if we wish to cut down 2·4647155 to five places of decimals, we should write 2·46472, whereas to cut down 2·4647145 we should write 2·46471.

14. Where i consists of more than one significant figure, we can generally find a short method of multiplication. Thus, if $i = \cdot 0425$, we can divide by 4, instead of multiplying by 25, of course correctly placing the result as regards the decimal point.

Again, if the rate be $4\frac{2}{3}$ per cent., we can divide by 6 the result of the multiplication by 4. A few lines of each of these examples are given in the margin.

	$(1+i)^n$ $4\frac{1}{4}$ per cent.
1.	1·0425
	41700
	2606
2.	1·086806
	43472
	2717
3.	1·132995
	45320
	2832
4.	1·181147
	47246
	2953
5.	1·231346
	* * *

	$(1+i)^n$ $4\frac{2}{3}$ per cent.
1.	1·046667
	41867
	6978
2.	1·095512
	43820
	7303
3.	1·146635
	45865
	7644
4.	1·200144
	48006
	8001
5.	1·256151
	* * *

15. Where i is such a number that multiplication becomes troublesome, recourse must be had to logarithms. This leads us to the next problem.

16. To construct a table of $\log (1+i)^n$.

Because $\log (1+i)^n = n \log (1+i)$, the table consists of the successive multiples of $\log (1+i)$, and these may be formed most conveniently by addition. The value of $\log (1+i)$ should be written at the top of the column and again at the foot of a card, which is moved down as the additions are performed. A verification is naturally obtained at every tenth value, the tenth being ten times the first, the twentieth ten times the second, etc.; the figures of each pair, therefore, being the same, the decimal point only being moved.

17. The last figure of $\log (1+i)$ is only approximately true, and the error in it is continuously multiplied as the work proceeds, so that a correction must be introduced to counteract the accumulation of error. Thus, at 4 per cent. $\log (1+i) = \cdot 01703334$. If in our operation we only retain six places of decimals, the value of $\log (1+i)^{10}$

will come out ·170330, and of $\log (1+i)^{100}$, 1·703300, whereas they should be ·170333 and 1·703334 respectively. We may keep correct the last place to be retained in one or other of two ways—either by working with two places more figures than are to appear in the final table, or by applying a correction as the work progresses. This second method is as follows:—The seventh and eighth figures of $\log (1+i)$, at 4 per cent., are 34, or almost exactly a third of a unit in the sixth place. If, therefore, we work with only six figures, and increase by a unit in the sixth place the second, fifth, eighth, etc., values, our results will be accurate. The specimen in the margin shows both these methods of correction. Eight figures are there used, although only six are to be retained, and the two that are to be cut off are separated by a space from the others. When the final cutting down process is effected, the usual correction must be made when the figures neglected are greater than 49.

$\log (1+i)^n$ 4 per cent.		
1.	017033	34
2.	034066	68
3.	051100	02
4.	068133	36
5.	085166	70
6.	102200	04
7.	119233	38
8.	136266	72
9.	153300	06
10.	170333	40
*	*	*

The necessity of going over the work to correct the last figure will be obviated if, at the top of the column, *but not on the moveable card*, we increase the seventh figure by 5. In the margin the operation is repeated with this adjustment, and it will be noticed that without further alteration the six figures to be retained are accurate.

$\log (1+i)^n$ 4 per cent.		
1.	017033	84
2.	034067	18
3.	051100	52
4.	068133	86
5.	085167	20
6.	102200	54
7.	119233	88
8.	136267	22
9.	153300	56
10.	170333	90
*	*	*

It will also be observed that if we had used only six places of figures in our operation, and, applying the second method of correction above named, if we had, to form the second, fifth, eighth, etc., values, added 017034, instead of 017033, which is used to form the other values, the result would have been also correct. By examining the first two figures of $\log (1+i)$ which are to be neglected, we can always see at what intervals in our work the last place which we retain is to be increased or diminished by a unit. Thus, at 6 per cent., $\log (1+i)$ is ·02530587, or, to six figures, ·025306. This last quantity is ·13 in excess in the last place, and therefore when we use it for continuous addition, we must diminish our results by a unit at the fourth value (because $4 \times \cdot 13 = \cdot 52 > \cdot 49$), and thereafter at every eighth value.

18. To construct a table of v^n.

Since $v^{n+1}=v\times v^n$, we could form the table by beginning with v, and multiplying continuously by v; but this would not be convenient. The quantity v has generally many significant figures, and the multiplications would therefore be lengthy. By changing the working formula into $v^n=(1+i)v^{n+1}$, we can reduce the labour, and make it no greater than that involved in preparing a table of $(1+i)^n$. Commencing then with the last value of v^n to be tabulated, we work backwards; but in other respects we proceed exactly as if we were preparing the column $(1+i)^n$. All that we have said regarding this last function will therefore apply, and it is unnecessary to repeat illustrations here.

19. It is very useful when we have completed an entire column of a table to be able to verify by one operation the whole work, or to be able at once to check a printed table with which we are not familiar. In the case of the majority of columns of interest tables, excellent formulas for this purpose are easily found. We have $s_{\overline{n}|}=1+(1+i)+(1+i)^2+\ .\ .\ .\ +(1+i)^{n-1}$. If therefore we add up our column of $(1+i)^n$ and increase the sum by unity, the result, if our work is correct, will be $s_{\overline{n+1}|}$. Similarly, $a_{\overline{n}|}=v+v^2+\ .\ .\ .\ v^n$, so that the sum of our column of v^n should be $a_{\overline{n}|}$.

20. To construct a table of $s_{\overline{n}|}$.

Since $s_{\overline{n+1}|}=s_{\overline{n}|}+(1+i)^n$, it follows that the column $(1+i)^n$ consists of the differences between the successive values in column $s_{\overline{n}|}$, and we can therefore, if we have already formed a table of $(1+i)^n$, construct a table of $s_{\overline{n}|}$ by mere summation. Commencing with 1, the amount of 1 per annum for one year, we add successively $(1+i)$, $(1+i)^2$, etc., so forming $s_{\overline{2}|}$, $s_{\overline{3}|}$, etc. At any stage the work may be checked by calculating independently the amount of the annuity by means of the formula $s_{\overline{n}|}=\frac{(1+i)^n-1}{i}$. As the values in the table of $(1+i)^n$ are only approximately true in the last place, the errors, although they are in both directions, and so will in general tend to neutralize each other, may sometimes at certain points in the table of $s_{\overline{n}|}$ fall in such a manner as to produce a sensible inaccuracy in the last place. We must therefore, to insure exactitude, work to at least one place more than we mean finally to retain.

21. If we do not already possess a table of $(1+i)^n$, we can nevertheless form our table of $s_{\overline{n|}}$ with great facility for the majority of rates of interest, by simple multiplication. The relation to be used is $s_{\overline{n+1|}}=(1+i)s_{\overline{n|}}+1$. At each step we multiply our previous result by $(1+i)$, exactly as we did in Arts. 13 and 14, the only difference in the operation being that now we add a unit at the same time that we make the multiplication. In the margin we show the construction of the table at 4 per cent. interest. It is needless to add further examples.

	$s_{\overline{n\vert}}$ 4 per cent.
1.	1·00
	1 4
2.	2·04
	1 816
3.	3·1216
	1·124864
4.	4·246464
	1·169859
5.	5·416323
	1·216653
6.	6·632976
	* * *

22. At each stage, for our addend we multiply our previous result by i and add unity: that is, to form $s_{\overline{n+1|}}$ we add to $s_{\overline{n|}}$ the quantity $i\,s_{\overline{n|}}+1$. But from the equation $s_{\overline{n|}}=\dfrac{(1+i)^n-1}{i}$ it follows that $i\,s_{\overline{n|}}+1=(1+i)^n$, and this shows, if proof be needed, that our addends are the successive powers of $(1+i)$. Therefore, in constructing our table of $s_{\overline{n|}}$ by this method, we at the same time form a table of $(1+i)^n$, and that without greater labour than when we construct the table of $(1+i)^n$ alone. It is thus desirable, even if we want only a table of $(1+i)^n$, to form—at least when the rate of interest is integral—the table of $s_{\overline{n|}}$, as by so doing we obtain two tables in one operation.

23. A very useful formula is available to check our complete column of $s_{\overline{n|}}$, or to verify a printed table of this function—

$$s_{\overline{1|}}+s_{\overline{2|}}+s_{\overline{3|}}+\text{etc.}+s_{\overline{n|}}$$
$$=\frac{(1+i)-1}{i}+\frac{(1+i)^2-1}{i}+\frac{(1+i)^3-1}{i}+\text{etc.}+\frac{(1+i)^n-1}{i}$$
$$=\frac{(1+i)+(1+i)^2+(1+i)^3+\text{etc.}+(1+i)^n-n}{i}=\frac{(1+i)s_{\overline{n|}}-n}{i}$$

If, therefore, we multiply the last value in the column by $(1+i)$, deduct n, and divide by i, we have as result the sum of the column. Should the equation hold we know that all our work is correct, unless indeed there be an exact balance of errors—an unlikely event; but, should the equation not hold, we may be sure that there is an error somewhere, and we must set to work to discover and eliminate it. We may divide our table into sections, and apply our formula again, and so localize the error. Thus, for example, the sum of the column under 4 per cent. in Table III. is 4687·75784, and, applying the formula, we find that $\dfrac{1{\cdot}04\times 237{\cdot}99069-60}{{\cdot}04}$ is

4687·75744, differing from the sum of the column by 40 in the last two places, which is due to the fact that the figures in the last place of the tabulated function are only approximately true, and that in multiplying $s_{\overline{60}|}$ by 1·04 we have not made allowance for the carriage from the figures beyond the fifth. Had we used six places of decimals in $s_{\overline{60}|}$, the result of the formula would have been 4687·75781, or only 3 out in the fifth place. Suppose, however, that the fiftieth value had by mistake been printed 153·66708, the result of addition would have differed from that given by the application of the formula by ·99960, thus showing that there must be a mistake somewhere. Splitting the column into three equal sections, we find the sum of the first twenty values to be 274·23005, while $\frac{(1+i)s_{\overline{20}|}-20}{i}=274{\cdot}23008$, thus showing that the error is not in that section. Summing the second section we have 1196·43339, which, added to the sum of the first section, gives 1470·66344, and $\frac{(1+i)s_{\overline{40}|}-40}{i}=1470{\cdot}66352$. The error is therefore not in the second section, and it must be in the last; and, examining the values in it, we find $s_{\overline{50}|}$ to be wrong.

24. To construct a table of $a_{\overline{n}|}$.

This table may be formed from a table of v^n exactly as a table of $s_{\overline{n}|}$ may be formed from a table of $(1+i)^n$.

If however we have not already a table of v^n, we can construct that of $a_{\overline{n}|}$ directly by means of the relation $a_{\overline{n-1}|}=(1+i)a_{\overline{n}|}-1$.

We must proceed as in Art. 21, except that we must begin at the end of the table and work backwards, and that we must deduct instead of add unity at each step. The example in the margin shows the construction at 4 per cent.

$a_{\overline{n}|}$ 4 per cent.	
60.	22·623490
	904940
59.	22·528430
	901137
58.	22·429567
	897183
57.	22·326750
	893070
56.	22·219820
	888793
55.	22·108613
	* * *

25. From the relation $a_{\overline{n}|}=\frac{1-v^n}{i}$ it follows that $i\,a_{\overline{n}|}=1-v^n$. Therefore the addends that we form are the arithmetical complements of v^n, from which v^n can be derived very easily, almost by inspection. Therefore, if we wish to construct a table of v^n, we may, without much additional labour, do so by first forming one of $a_{\overline{n}|}$, and so secure both tables at once.

26. A very similar relation to that given in Art. 23 is available to check our column of $a_{\overline{n}|}$. We have

$$a_{\overline{1}|}+a_{\overline{2}|}+a_{\overline{3}|}+\text{etc.}+a_{\overline{n}|}$$
$$=\frac{1-v}{i}+\frac{1-v^2}{i}+\frac{1-v^3}{i}+\text{etc.}+\frac{1-v^n}{i}=\frac{n-a_{\overline{n}|}}{i}.$$

This formula may be applied exactly as in Art. 23 to detect errors.

27. We may here mention that for tables of all kinds a very useful check can be applied by simply differencing. The differences of a table almost always follow some sufficiently defined law to render apparent any irregularities produced by errors. Thus, in the case of the supposititious error discussed in Art. 23, we should find the differences of the column at that point to run as follows:—

47.	6·31782
48.	6·57052
49.	7·83335
50.	6·10669
51.	7·39095
52.	7·68659

The difference between the forty-ninth and fiftieth values is evidently too great, and that between the fiftieth and fifty-first too small, and it is thus seen that the fiftieth value is wrong. If we make the proper correction, the differences will run smoothly. It will very often happen that the differences need not be actually taken out and entered on paper. The operation can be performed mentally by a careful inspection of the table.

28. Logarithms may very conveniently be employed to construct tables of $s_{\overline{n}|}$, and $a_{\overline{n}|}$, if they are arranged in the form first given by Gauss. In Gauss's logarithmic table the number with which the table is entered, called the "argument," is $\log x$, and the result is $\log(1+x)$. If by the letter T prefixed to a quantity, we denote the result of entering a table with that quantity, then the property above mentioned of Gauss's tables may be represented by the equation $T\log x=\log(1+x)$.

Mr. Peter Gray was the first to apply, in his book, *Tables and Formulæ for the Computation of Life Contingencies*, Gauss's logarithmic tables to the calculation of life annuities and assurances, and of annuities-certain, and for that purpose he recomputed and extended Gauss's tables. In Mr. Gray's tables we have $\log(1+x)$ given to six places of decimals for all values of $\log x$, from $x=\cdot001$ to $x=100$. Since Mr. Gray's tables were issued, Wittstein's more extensive ones of the same kind have been published. These give to seven places $\log(1+x)$ for all values of x, from $x=\cdot0000001$ to $x=1000000$. A small five-place table has also been published by Messrs. Galbraith & Haughton giving to five places $\log(1+x)$ for all values of x, from $x=1$ to $x=10000$.

29. To construct by Gauss's logarithms a table of $s_{\overline{n}|}$,

We have $s_{\overline{n}|}=(1+i)\,s_{\overline{n-1}|}+1$; whence

$$\log s_{\overline{n}|}=\log\{(1+i)\,s_{\overline{n-1}|}+1\}.$$

Now, if we know $\log\{(1+i)\,s_{\overline{n-1}|}\}$, that is $\{\log(1+i)+\log s_{\overline{n-1}|}\}$, we can, by Gauss's tables, without finding the natural numbers, obtain $\log\{(1+i)\,s_{\overline{n-1}|}+1\}$. We have only to enter the table with the sum of the two given logarithms, and the result is the logarithm which we require. The equation may therefore be written

$$\log s_{\overline{n}|}=T\{\log s_{\overline{n-1}|}+\log(1+i)\}.$$

We commence with $\log s_{\overline{1}|}$, which is 0 since $s_{\overline{1}|}=1$, and write below it, and on every succeeding third line, the constant quantity $\log(1+i)$. Adding the first two lines together, and placing the sum on the third line, we have $\{\log s_{\overline{1}|}+\log(1+i)\}$, with which we enter the table, and have for result $\log s_{\overline{2}|}$, which we place on the fourth line. Adding to this, again, $\log(1+i)$, which we have already placed on the fifth line, and entering the table with the sum, we have $\log s_{\overline{3}|}$, and so on. In the example in the margin, a few values of $\log s_{\overline{n}|}$, at 4 per cent., are worked out, and Mr. Gray's tables of Gaussian logarithms are used. If we write once for all $\log(1+i)$ at the foot of a card, to be moved down as the work proceeds, we shall save ourselves the trouble of repeating it on every third line, and our calculations will not occupy so much space. Seeing that the seventh figure of $\log(1+i)$, at 4 per cent., is 3—or one-third of a unit in the sixth place—we must, at every third value, increase by a unit the sixth figure. This we have done in the example.

$\log s_{\overline{n}|}$ 4 per cent.	
$\log s_{\overline{1}|}$	·000000
$\log(1+i)$	·017033
	·017033
$\log s_{\overline{2}|}$	·309630
$\log(1+i)$	·017033
	·326663
$\log s_{\overline{3}|}$	·494377
$\log(1+i)$	·017034
	·511411
$\log s_{\overline{4}|}$	·628027
$\log(1+i)$	·017033
	·645060
$\log s_{\overline{5}|}$	·733704
$\log(1+i)$	·017033
	* * *

When $s_{\overline{n}|}$ becomes greater than 100, Mr. Gray's Gaussian logarithms are no longer available, and the table must be completed in some other way. Mr. Gray shows how to do so by means of a table of $\log(1-x)$, which he also gives, and those who wish to pursue the subject further may consult his book. With Wittstein's and Galbraith & Haughton's editions of Gauss's logarithms the difficulty does not arise.

30. To construct by Gauss's logarithms a table of $a_{\overline{n}|}$.

From the equation $a_{\overline{n}|}=v(1+a_{\overline{n-1}|})$ we pass to $\log a_{\overline{n}|}=\log v+\log(1+a_{\overline{n-1}|})$, which may be written

$$\log a_{\overline{n}|}=\log v+T\log a_{\overline{n-1}|}.$$

As each value of $\log a$ is formed, we enter the table with it, and to the result add $\log v$, thus obtaining the next value of $\log a$. The example in the margin gives a few lines of the work at 4 per cent. Because we continuously add $\log v$, we must make the usual correction for the last place as we proceed. At 4 per cent. $\log v$ is ·3 in excess in the sixth place, and we must therefore deduct a unit every third time it is used. As before, we may write $\log v$ on a moveable card to save trouble.

$\log a_{\overline{n}|}$ 4 per cent.	
$\log a_{\overline{1}|} = \log v$	982967
$T\log a_{\overline{1}|}$	292597
$\log a_{\overline{2}|}$	275564
$\log v$	982967
$T\log a_{\overline{2}|}$	460311
$\log a_{\overline{3}|}$	443278
$\log v$	982966
$T\log a_{\overline{3}|}$	576928
$\log a_{\overline{4}|}$	559894
$\log v$	982967
$T\log a_{\overline{4}|}$	665571
$\log a_{\overline{5}|}$	648538
	* * *

31. There is no continuous formula for the construction of the columns $s^{-1}_{\overline{n}|}$ and $a^{-1}_{\overline{n}|}$. We must take the reciprocals of $s_{\overline{n}|}$ and $a_{\overline{n}|}$ respectively. In order to insure accuracy, it will therefore be necessary carefully to verify *each* tabulated value. This may best be done, not by performing the work in duplicate, as the same error might thus be repeated, but by again taking out the reciprocals of the values in the newly formed columns, and these should be the values of $s_{\overline{n}|}$ and $a_{\overline{n}|}$ respectively, with which we started.

32. If both the columns $s^{-1}_{\overline{n}|}$ and $a^{-1}_{\overline{n}|}$ are wanted, one may be formed from the other in such a way as to check them both. Since $s^{-1}_{\overline{n}|}=a^{-1}_{\overline{n}|}-i$, it follows that the two columns have the same differences. If, therefore, we first form the column $s^{-1}_{\overline{n}|}$ by means of a table of reciprocals, and difference it, we shall so produce the differences (which are negative) of the column $a^{-1}_{\overline{n}|}$ Starting now with the first value of the column $a^{-1}_{\overline{n}|}$, namely, $(1+i)$, and adding (algebraically) continuously the differences, we complete the column. To check the whole work, we take the reciprocals of the values in the last column, and these should be the successive values of $a_{\overline{n}|}$.

The following is an example at 4 per cent.

Year	$s_{\overline{n}\vert}^{-1}$	$-\Delta$	$a_{\overline{n}\vert}^{-1}$
1	1·000000	490196	1·040000
2	·490196	830153	·530196
3	·320349	915141	·360349
4	·235490	949137	·275490
5	·184627	966135	·224627
6	·150762	975848	·190762
7	·126610	981918	·166610
8	·108528	985965	·148528
9	·094493	988798	·134493
10	·083291		·123291

Where i consists of but one significant figure, as in the example, this method of construction does not possess any advantages; but if there be several significant figures in i, then considerable benefits are experienced by its adoption.

33. The column $\log s_{\overline{n}|}^{-1}$ may be conveniently formed by means of the logarithms of s_{n}, because, omitting the characteristics of the logarithms, $\log s_{\overline{n}|}^{-1} = 1 - \log s_{\overline{n}}$. We have only to take the arithmetical complements of the logarithms of the successive values of $s_{\overline{n}|}$, which can be done by inspection. In the same way $\log a_{\overline{n}|}^{-1}$ may be formed from $\log a_{\overline{n}}$.

3	1·092727	1·108718	1·124864	1·141166	1·157625	1·191016	3
4	1·125509	1·147523	1·169859	1·192519	1·215506	1·262477	4
5	1·159274	1·187686	1 216653	1·246182	1·276282	1·338226	5
6	1·194052	1·229255	1·265319	1·302260	1·340096	1·418519	6
7	1·229874	1·272279	1·315932	1·360862	1·407100	1·503630	7
8	1·266770	1·316809	1·368569	1·422101	1·477455	1·593848	8
9	1·304773	1·362897	1·423312	1·486095	1·551328	1 689479	9
10	1·343916	1·410599	1·480244	1·552969	1·628895	1·790848	10
1	1·384234	1·459970	1·539454	1·622853	1·710339	1·898299	1
2	1·425761	1·511069	1·601032	1·695881	1·795856	2·012196	2
3	1·468534	1·563956	1·665074	1·772196	1·885649	2·132928	3
4	1·512590	1·618695	1·731676	1·851945	1·979932	2·260904	4
15	1·557967	1·675349	1·800944	1·935282	2·078928	2·396558	15
6	1·604706	1·733986	1·872981	2·022370	2·182875	2·540352	6
7	1·652848	1·794676	1·947901	2·113377	2·292018	2·692773	7
8	1·702433	1·857489	2·025816	2·208479	2·406619	2·854339	8
9	1·753506	1·922501	2·106849	2·307860	2·526950	3·025600	9
20	1·806111	1·989789	2·191123	2·411714	2·653298	3·207135	20
1	1·860295	2·059431	2·278768	2·520241	2·785963	3·399564	1
2	1·916103	2·131512	2·369919	2·633652	2·925261	3·603537	2
3	1·973587	2·206114	2·464716	2·752166	3·071524	3·819750	3
4	2·032794	2·283328	2·563304	2·876014	3·225100	4·048935	4
25	2·093778	2·363245	2·665836	3·005434	3·386355	4·291871	25
6	2·156591	2·445959	2·772470	3·140679	3·555673	4·549383	6
7	2·221289	2·531567	2·883369	3·282010	3·733456	4·822346	7
8	2·287928	2·620172	2·998703	3·429700	3·920129	5·111687	8
9	2·356566	2·711878	3·118651	3·584036	4·116136	5·418388	9
30	2·427262	2·806794	3·243398	3·745318	4·321942	5·743491	30
1	2·500080	2·905031	3·373133	3·913857	4·538039	6·088101	1
2	2·575083	3·006708	3·508059	4·089981	4·764941	6·453387	2
3	2·652335	3·111942	3·648381	4·274030	5·003189	6·840590	3
4	2·731905	3·220860	3·794316	4·466362	5·253348	7·251025	4
35	2·813862	3·333590	3·946089	4·667348	5·516015	7·686087	35
6	2·898278	3·450266	4·103933	4·877378	5·791816	8·147252	6
7	2·985227	3·571025	4·268090	5·096860	6·081407	8·636087	7
8	3·074783	3·696011	4·438813	5·326219	6·385477	9·154252	8
9	3·167027	3·825372	4·616366	5·565899	6·704751	9·703508	9
40	3·262038	3·959260	4·801021	5·816365	7·039989	10·285718	40
1	3·359899	4·097834	4·993061	6·078101	7·391988	10·902861	1
2	3·460696	4·241258	5·192784	6·351615	7·761588	11·557033	2
3	3·564517	4·389702	5·400495	6·637438	8·149667	12·250454	3
4	3·671452	4·543342	5·616515	6·936123	8·557150	12·985482	4
45	3·781596	4·702359	5·841176	7·248248	8·985008	13·764611	45
6	3·895044	4·866941	6·074823	7·574420	9·434258	14·590487	6
7	4·011895	5·037284	6·317816	7·915268	9·905971	15·465917	7
8	4·132252	5·213589	6·570528	8·271456	10·401270	16·393872	8
9	4·256219	5·396065	6·833349	8·643671	10·921333	17·377504	9
50	4·383906	5.584927	7·106683	9·032636	11·467400	18·420154	50
1	4·515423	5·780399	7·390951	9·439105	12·040770	19·525364	1
2	4·650886	5·982713	7·686589	9·863865	12·642808	20·696885	2
3	4·790412	6·192108	7·994052	10·307739	13·274949	21·938698	3
4	4·934125	6·408832	8·313814	10·771587	13·938696	23·255020	4
55	5·082149	6·633141	8·646367	11·256308	14·635631	24·650322	55
6	5·234613	6·865301	8·992222	11·762842	15·367412	26·129341	6
7	5·391651	7·105587	9·351910	12·292170	16·135783	27·697101	7
8	5·553401	7·354282	9·725987	12·845318	16·942572	29·358927	8
9	5·720003	7·611682	10·115026	13·423357	17·789701	31·120463	9
60	5·891603	7·878091	10·519627	14·027408	18·679186	32·987691	60

Present value of 1 :—*viz.* v^n.

n	3%	$3\frac{1}{2}$%	4%	$4\frac{1}{2}$%	5%	6%	n
1	·970874	·966184	·961538	·956938	·952381	·943396	1
2	·942596	·933511	·924556	·915730	·907029	·889996	2
3	·915142	·901943	·888996	·876297	·863838	·839619	3
4	·888487	·871442	·854804	·838561	·822702	·792094	4
5	·862609	·841973	·821927	·802451	·783526	·747258	5
6	·837484	·813501	·790315	·767896	·746215	·704961	6
7	·813092	·785991	·759918	·734828	·710681	·665057	7
8	·789409	·759412	·730690	·703185	·676839	·627412	8
9	·766417	·733731	·702587	·672904	·644609	·591898	9
10	·744094	·708919	·675564	·643928	·613913	·558395	10
1	·722421	·684946	·649581	·616199	·584679	·526788	1
2	·701380	·661783	·624597	·589664	·556837	·496969	2
3	·680951	·639404	·600574	·564272	·530321	·468839	3
4	·661118	·617782	·577475	·539973	·505068	·442301	4
15	·641862	·596891	·555265	·516720	·481017	·417265	15
6	·623167	·576706	·533908	·494469	·458112	·393646	6
7	·605016	·557204	·513373	·473176	·436297	·371364	7
8	·587395	·538361	·493628	·452800	·415521	·350344	8
9	·570286	·520156	·474642	·433302	·395734	·330513	9
20	·553676	·502566	·456387	·414643	·376889	·311805	20
1	·537549	·485571	·438834	·396787	·358942	·294155	1
2	·521893	·469151	·421955	·379701	·341850	·277505	2
3	·506692	·453286	·405726	·363350	·325571	·261797	3
4	·491934	·437957	·390121	·347703	·310068	·246979	4
25	·477606	·423147	·375117	·332731	·295303	·232999	25
6	·463695	·408838	·360689	·318402	·281241	·219810	6
7	·450189	·395012	·346817	·304691	·267848	·207368	7
8	·437077	·381654	·333477	·291571	·255094	·195630	8
9	·424346	·368748	·320651	·279015	·242946	·184557	9
30	·411987	·356278	·308319	·267000	·231377	·174110	30
1	·399987	·344230	·296460	·255502	·220359	·164255	1
2	·388337	·332590	·285058	·244500	·209866	·154957	2
3	·377026	·321343	·274094	·233971	·199873	·146186	3
4	·366045	·310476	·263552	·223896	·190355	·137912	4
35	·355383	·299977	·253415	·214254	·181290	·130105	35
6	·345032	·289833	·243669	·205028	·172657	·122741	6
7	·334983	·280032	·234297	·196199	·164436	·115793	7
8	·325226	·270562	·225285	·187750	·156605	·109239	8
9	·315754	·261413	·216621	·179665	·149148	·103056	9
40	·306557	·252572	·208289	·171929	·142046	·097222	40
1	·297628	·244031	·200278	·164525	·135282	·091719	1
2	·288959	·235779	·192575	·157440	·128840	·086527	2
3	·280543	·227806	·185168	·150661	·122704	·081630	3
4	·272372	·220102	·178046	·144173	·116861	·077009	4
45	·264439	·212659	·171198	·137964	·111297	·072650	45
6	·256737	·205468	·164614	·132023	·105997	·068538	6
7	·249259	·198520	·158283	·126338	·100949	·064658	7
8	·241999	·191806	·152195	·120898	·096142	·060998	8
9	·234950	·185320	·146341	·115692	·091564	·057546	9
50	·228107	·179053	·140713	·110710	·087204	·054288	50
1	·221463	·172998	·135301	·105942	·083051	·051215	1
2	·215013	·167148	·130097	·101380	·079096	·048316	2
3	·208750	·161496	·125093	·097014	·075330	·045582	3
4	·202670	·156035	·120282	·092837	·071743	·043001	4
55	·196767	·150758	·115656	·088839	·068326	·040567	55
6	·191036	·145660	·111207	·085013	·065073	·038271	6
7	·185472	·140734	·106930	·081353	·061974	·036105	7
8	·180070	·135975	·102817	·077849	·059023	·034061	8
9	·174825	·131377	·098863	·074497	·056212	·032133	9
60	·169733	·126934	·095060	·071289	·053536	·030314	60

Table III.

Amount of 1 per annum:—viz. $s_{\overline{n}|}$.

n	3%	3½%	4%	4½%	5%	6%	n
1	1.00000	1.00000	1.00000	1.00000	1.00000	1.00000	1
2	2.03000	2.03500	2.04000	2.04500	2.05000	2.06000	2
3	3.09090	3.10623	3.12160	3.13703	3.15250	3.18360	3
4	4.18363	4.21494	4.24646	4.27819	4.31013	4.37462	4
5	5.30914	5.36247	5.41632	5.47071	5.52563	5.63709	5
6	6.46841	6.55015	6.63298	6.71689	6.80191	6.97532	6
7	7.66246	7.77941	7.89829	8.01915	8.14201	8.39384	7
8	8.89234	9.05169	9.21423	9.38001	9.54911	9.89747	8
9	10.15911	10.36850	10.58280	10.80211	11.02656	11.49132	9
10	11.46388	11.73139	12.00611	12.28821	12.57789	13.18080	10
1	12.80780	13.14199	13.48635	13.84118	14.20679	14.97164	1
2	14.19203	14.60196	15.02581	15.46403	15.91713	16.86994	2
3	15.61779	16.11303	16.62684	17.15991	17.71298	18.88214	3
4	17.08632	17.67699	18.29191	18.93211	19.59863	21.01507	4
15	18.59891	19.29568	20.02359	20.78405	21.57856	23.27597	15
6	20.15688	20.97103	21.82453	22.71934	23.65749	25.67253	6
7	21.76159	22.70502	23.69751	24.74171	25.84037	28.21288	7
8	23.41444	24.49969	25.64541	26.85508	28.13239	30.90565	8
9	25.11687	26.35718	27.67123	29.06356	30.53900	33.75999	9
20	26.87037	28.27968	29.77808	31.37142	33.06595	36.78559	20
1	28.67649	30.26947	31.96920	33.78314	35.71925	39.99273	1
2	30.53678	32.32890	34.24797	36.30338	38.50521	43.39229	2
3	32.45288	34.46041	36.61789	38.93703	41.43048	46.99583	3
4	34.42647	36.66653	39.08260	41.68920	44.50200	50.81558	4
25	36.45926	38.94986	41.64591	44.56521	47.72710	54.86451	25
6	38.55304	41.31310	44.31174	47.57065	51.11345	59.15638	6
7	40.70963	43.75906	47.08421	50.71132	54.66913	63.70577	7
8	42.93092	46.29063	49.96758	53.99333	58.40259	68.52811	8
9	45.21885	48.91080	52.96629	57.42303	62.32271	73.63980	9
30	47.57542	51.62268	56.08494	61.00707	66.43885	79.05819	30
1	50.00268	54.42947	59.32834	64.75239	70.76079	84.80168	1
2	52.50276	57.33450	62.70147	68.66625	75.29883	90.88978	2
3	55.07784	60.34121	66.20953	72.75623	80.06377	97.34317	3
4	57.73018	63.45315	69.85791	77.03026	85.06696	104.18376	4
35	60.46208	66.67401	73.65222	81.49662	90.32031	111.43478	35
6	63.27594	70.00760	77.59831	86.16397	95.83632	119.12087	6
7	66.17422	73.45787	81.70225	91.04134	101.62814	127.26812	7
8	69.15945	77.02889	85.97034	96.13821	107.70955	135.90421	8
9	72.23423	80.72491	90.40915	101.46442	114.09502	145.05846	9
40	75.40126	84.55028	95.02552	107.03032	120.79977	154.76197	40
1	78.66330	88.50954	99.82654	112.84669	127.83976	165.04768	1
2	82.02320	92.60737	104.81960	118.92479	135.23175	175.95055	2
3	85.48389	96.84863	110.01238	125.27640	142.99334	187.50758	3
4	89.04841	101.23833	115.41288	131.91384	151.14301	199.75803	4
45	92.71986	105.78167	121.02939	138.84997	159.70016	212.74351	45
6	96.50146	110.48403	126.87057	146.09821	168.68516	226.50813	6
7	100.39650	115.35097	132.94539	153.67263	178.11942	241.09861	7
8	104.40840	120.38826	139.26321	161.58790	188.02539	256.56453	8
9	108.54065	125.60185	145.83373	169.85936	198.42666	272.95840	9
50	112.79687	130.99791	152.66708	178.50303	209.34800	290.33591	50
1	117.18077	136.58284	159.77377	187.53567	220.81540	308.75606	1
2	121.69620	142.36324	167.16472	196.97477	232.85617	328.28142	2
3	126.34708	148.34595	174.85131	206.83863	245.49897	348.97831	3
4	131.13749	154.53806	182.84536	217.14637	258.77392	370.91701	4
55	136.07162	160.94689	191.15917	227.91796	272.71262	394.17203	55
6	141.15377	167.58003	199.80554	239.17427	287.34825	418.82235	6
7	146.38838	174.44533	208.79776	250.93711	302.71566	444.95169	7
8	151.78003	181.55092	218.14967	263.22928	318.85145	472.64879	8
9	157.33343	188.90520	227.87566	276.07460	335.79402	502.00772	9
60	163.05344	196.51688	237.99069	289.49796	353.58372	533.12818	60

TABLE IV.

Present Value of 1 per annum:—*viz.* $a_{\overline{n}|}$

n	3%	3½%	4%	4½%	5%	6%	n
1	·97087	·96618	·96154	·95694	·95238	·94340	1
2	1·91347	1·89969	1·88609	1·87267	1·85941	1·83339	2
3	2·82861	2·80164	2·77509	2·74896	2·72325	2·67301	3
4	3·71710	3·67308	3·62990	3·58753	3·54595	3·46511	4
5	4·57971	4·51505	4·45182	4·38998	4·32948	4·21236	5
6	5·41719	5·32855	5·24214	5·15787	5·07569	4·91732	6
7	6·23028	6·11454	6·00205	5·89270	5·78637	5·58238	7
8	7·01969	6·87396	6·73275	6·59589	6·46321	6·20979	8
9	7·78611	7·60769	7·43533	7·26879	7·10782	6·80169	9
10	8·53020	8·31661	8·11090	7·91272	7·72174	7·36009	10
1	9·25262	9·00155	8·76048	8·52892	8·30641	7·88688	1
2	9·95400	9·66333	9·38507	9·11858	8·86325	8·38384	2
3	10·63496	10·30274	9·98565	9·68285	9·39357	8·85268	3
4	11·29607	10·92052	10·56312	10·22283	9·89864	9·29498	4
15	11·93794	11·51741	11·11839	10·73955	10·37966	9·71225	15
6	12·56110	12·09412	11·65230	11·23402	10·83777	10·10590	6
7	13·16612	12·65132	12·16567	11·70719	11·27407	10·47726	7
8	13·75351	13·18968	12·65930	12·15999	11·68959	10·82760	8
9	14·32380	13·70984	13·13394	12·59329	12·08532	11·15812	9
20	14·87748	14·21240	13·59033	13·00794	12·46221	11·46992	20
1	15·41502	14·69797	14·02916	13·40472	12·82115	11·76408	1
2	15·93692	15·16713	14·45112	13·78443	13·16300	12·04158	2
3	16·44361	15·62041	14·85684	14·14778	13·48857	12·30338	3
4	16·93554	16·05837	15·24696	14·49548	13·79864	12·55036	4
25	17·41315	16·48152	15·62208	14·82821	14·09395	12·78336	25
6	17·87684	16·89035	15·98277	15·14661	14·37519	13·00317	6
7	18·32703	17·28537	16·32959	15·45130	14·64303	13·21053	7
8	18·76411	17·66702	16·66306	15·74287	14·89813	13·40616	8
9	19·18846	18·03577	16·98372	16·02189	15·14107	13·59072	9
30	19·60044	18·39205	17·29203	16·28889	15·37245	13·76483	30
1	20·00043	18·73628	17·58849	16·54439	15·59281	13·92909	1
2	20·38877	19·06887	17·87355	16·78889	15·80268	14·08404	2
3	20·76579	19·39021	18·14765	17·02286	16·00255	14·23023	3
4	21·13184	19·70068	18·41120	17·24676	16·19290	14·36814	4
35	21·48722	20·00066	18·66461	17·46101	16·37419	14·49825	35
6	21·83225	20·29049	18·90828	17·66604	16·54685	14·62099	6
7	22·16724	20·57053	19·14258	17·86224	16·71129	14·73678	7
8	22·49246	20·84109	19·36786	18·04999	16·86789	14·84602	8
9	22·80822	21·10250	19·58449	18·22966	17·01704	14·94908	9
40	23·11477	21·35507	19·79277	18·40158	17·15909	15·04630	40
1	23·41240	21·59910	19·99305	18·56611	17·29437	15·13802	1
2	23·70136	21·83488	20·18563	18·72355	17·42321	15·22454	2
3	23·98190	22·06269	20·37080	18·87421	17·54591	15·30617	3
4	24·25427	22·28279	20·54884	19·01838	17·66277	15·38318	4
45	24·51871	22·49545	20·72004	19·15635	17·77407	15·45583	45
6	24·77545	22·70092	20·88465	19·28837	17·88007	15·52437	6
7	25·02471	22·89944	21·04294	19·41471	17·98102	15·58903	7
8	25·26671	23·09124	21·19513	19·53561	18·07716	15·65003	8
9	25·50166	23·27656	21·34147	19·65130	18·16872	15·70757	9
50	25·72976	23·45562	21·48219	19·76201	18·25593	15·76186	50
1	25·95123	23·62862	21·61749	19·86795	18·33898	15·81308	1
2	26·16624	23·79577	21·74758	19·96933	18·41807	15·86139	2
3	26·37499	23·95726	21·87268	20·06635	18·49340	15·90697	3
4	26·57766	24·11330	21·99296	20·15918	18·56515	15·94998	4
55	26·77443	24·26405	22·10861	20·24802	18·63347	15·99054	55
6	26·96546	24·40971	22·21982	20·33303	18·69855	16·02881	6
7	27·15094	24·55045	22·32675	20·41439	18·76052	16·06492	7
8	27·33101	24·68642	22·42957	20·49224	18·81954	16·09898	8
9	27·50583	24·81780	22·52843	20·56673	18·87575	16·13111	9
60	27·67556	24·94473	22·62349	20·63802	18·92929	16·16143	60

TABLE V.

Annuity which 1 will purchase:—viz. $(a_{\overline{n}|})^{-1}$.

n	3%	3½%	4%	4½%	5%	6%	n
1	1·030000	1·035000	1·040000	1·045000	1·050000	1·060000	1
2	0·522611	0·526400	0·530196	·533998	·537805	0·545437	2
3	·353530	·356934	·360349	·363773	·367209	·374110	3
4	·269027	·272251	·275490	·278744	·282012	·288591	4
5	·218355	·221481	·224627	·227792	·230975	·237396	5
6	·184598	·187668	·190762	·193878	·197017	·203363	6
7	·160506	·163544	·166610	·169701	·172820	·179135	7
8	·142456	·145477	·148528	·151610	·154722	·161036	8
9	·128434	·131446	·134493	·137574	·140690	·147022	9
10	·117231	·120241	·123291	·126379	·129505	·135868	10
1	·108077	·111092	·114149	·117248	·120389	·126793	1
2	·100462	·103484	·106552	·109666	·112825	·119277	2
3	·094030	·097062	·100144	·103275	·106456	·112960	3
4	·088526	·091571	·094669	·097820	·101024	·107585	4
15	·083767	·086825	·089941	·093114	·096342	·102963	15
6	·079611	·082685	·085820	·089015	·092270	·098952	6
7	·075953	·079043	·082199	·085418	·088699	·095445	7
8	·072709	·075817	·078993	·082237	·085546	·092357	8
9	·069814	·072940	·076139	·079407	·082745	·089621	9
20	·067216	·070361	·073582	·076876	·080243	·087185	20
1	·064872	·068037	·071280	·074601	·077996	·085005	1
2	·062747	·065932	·069199	·072546	·075971	·083046	2
3	·060814	·064019	·067309	·070682	·074137	·081278	3
4	·059047	·062273	·065587	·068987	·072471	·079679	4
25	·057428	·060674	·064012	·067439	·070952	·078227	25
6	·055938	·059205	·062567	·066021	·069564	·076904	6
7	·054564	·057852	·061239	·064719	·068292	·075697	7
8	·053293	·056603	·060013	·063521	·067123	·074593	8
9	·052115	·055445	·058880	·062415	·066046	·073580	9
30	·051019	·054371	·057830	·061392	·065051	·072649	30
1	·049999	·053372	·056855	·060443	·064132	·071792	1
2	·049047	·052442	·055949	·059563	·063280	·071002	2
3	·048156	·051572	·055104	·058745	·062490	·070273	3
4	·047322	·050760	·054315	·057982	·061755	·069598	4
35	·046539	·049998	·053577	·057270	·061072	·068974	35
6	·045804	·049284	·052887	·056606	·060434	·068395	6
7	·045112	·048613	·052240	·055984	·059840	·067857	7
8	·044459	·047982	·051632	·055402	·059284	·067358	8
9	·043844	·047388	·051061	·054855	·058765	·066894	9
40	·043262	·046827	·050523	·054343	·058278	·066462	40
1	·042712	·046298	·050017	·053862	·057822	·066059	1
2	·042192	·045798	·049540	·053409	·057395	·065683	2
3	·041698	·045325	·049090	·052982	·056993	·065333	3
4	·041230	·044878	·048665	·052581	·056616	·065006	4
45	·040785	·044453	·048262	·052202	·056262	·064701	45
6	·040363	·044051	·047882	·051845	·055928	·064415	6
7	·039961	·043669	·047522	·051507	·055614	·064148	7
8	·039578	·043306	·047181	·051189	·055318	·063898	8
9	·039213	·042962	·046857	·050887	·055040	·063664	9
50	·038865	·042634	·046550	·050602	·054777	·063444	50
1	·038534	·042322	·046259	·050332	·054529	·063239	1
2	·038217	·042024	·045982	·050077	·054294	·063046	2
3	·037915	·041741	·045719	·049835	·054073	·062866	3
4	·037626	·041471	·045469	·049605	·053864	·062696	4
55	·037349	·041213	·045231	·049388	·053667	·062537	55
6	·037084	·040967	·045005	·049181	·053480	·062388	6
7	·036831	·040732	·044789	·048985	·053303	·062247	7
8	·036588	·040508	·044584	·048799	·053136	·062116	8
9	·036356	·040294	·044388	·048622	·052978	·061992	9
60	·036133	·040089	·044202	·048454	·052828	·061876	60

www.ingramcontent.com/pod-product-compliance
Lightning Source LLC
Chambersburg PA
CBHW021828190326
41518CB00007B/787